AF597351

CREATING THE HIGH PERFORMANCE INTERNATIONAL PETROLEUM COMPANY:

DINOSAURS CAN FLY

ETIENNE DEFFARGES
PHIL ELLIS
AL ESCHER
CYRUS FRIEDHEIM
TIM LASETER
JAY MARSHALL
NEIL McARTHUR
MATT McKENNA
BAXTOR NAIRON
BRUCE PASTERNACK
HARRY QUARLS
JOE QUOYSER
MATT ROGERS
BERTRAND SHELTON
ERIC SPIEGEL
RICHARD SPITZER
EARLE STEINBERG
DOUWE TIDEMAN
ROSS TOKMAKIAN
ALBERT VISCIO

JOHN ELTING TREAT, EDITOR

PennWell Books

PENNWELL PUBLISHING COMPANY
TULSA • OKLAHOMA

CREATING THE HIGH PERFORMANCE INTERNATIONAL PETROLEUM COMPANY:

DINOSAURS CAN FLY

PennWell Publishing Company
1421 South Sheridan/P.O. Box 1260
Tulsa, Oklahoma 74101

Library of Congress Cataloging in Publication Data

Treat, John Elting, et. al.
Creating the high performance international petroleum company: dinosaurs can fly / By John Elting Treat, et. al.
p. cm.
Includes bibliographical references and index.
ISBN 0-87814-429-3
1. Petroleum industry and trade—Management I. Title
HD9560.5.T68 1994
622'.338'068-dc20 94-23727 CIP

Printed in the United States of America

CONTENTS

AUTHORS' BIOGRAPHIES

Etienne Deffarges is a vice president and partner in the Paris office of Booz·Allen & Hamilton and codirector of the firm's European energy practice. He specializes in strategy development and organizational analysis for international oil and gas companies. He has led numerous international energy assignments for American, European, and Latin American companies, as well as several governments. Prior to joining Booz·Allen, he was a general field engineer with Schlumberger's wireline organization.

Mr. Deffarges graduated with high distinction as a Baker Scholar from the MBA program at Harvard Business School. He holds an MS in civil engineering from the University of California at Berkeley and a BS/MS in aeronautical engineering from the French National School of Aeronautics and Space in Toulouse, France.

Phillip Ellis is a Vice President of Booz·Allen and Hamilton and is codirector of the Energy Practice in Europe. Mr. Ellis has advised major oil and gas and electric power companies in Europe, the United States, and Asia on issues of strategy, organization design and change management for fifteen years. He is a contributor to the *Oil and Gas Journal*. Before joining Booz·Allen, Mr. Ellis was president of Miller/Ellis Oil Company and was previously with McKinsey and Company. Mr. Ellis received an MS in management from the Sloan School of Management at the Massachusetts Institute of Technology, where he was associated with the MIT Energy Laboratory. He also received a MA in international affairs from the George Washington University and a BS in physics from the University of California.

Al Escher is a principal in the energy and chemicals group of Booz·Allen, specializing in strategy development, organization design and implementation for energy companies. Mr. Escher has managed numerous projects both domestically and internationally in corporate goal setting and strategic planning, marketing strategy and implementation, organization design and implementation, and technology development. Prior to joining Booz·Allen & Hamilton, he spent six years with Texaco, in a variety of engineering, planning, and M&A assignments.

Mr. Escher received a BSE from Princeton University, an MS from the University of Southern California, and an MM from the J.L. Kellogg

Graduate School of Management, where he was elected to Beta Gamma Sigma for finishing in the top 10% of his class.

Cyrus Freidheim is vice chairman of Booz·Allen & Hamilton. He has 27 years of consulting experience in strategic planning, organization, marketing, finance and control, and information systems. In recent years, Mr. Freidheim has focused on strategy and restructuring for large, multinational corporations in the energy, automotive, consumer product, airline, and media industries. Mr. Freidheim has written and spoken on corporate restructuring and governance. He has researched and consulted on the evolution of the global corporation. Mr. Freidheim is a board member of Household International, Security Capital Group, and LaSalle Street Fund. He is a member of the America China Society, the U.S.-Japan Business Council, the Council on Foreign Relations, and sits on the boards of several civic, cultural, and educational institutions.

Tim Laseter is a principal in the Operations Management Group of Booz·Allen, specializing in the area of strategic sourcing. In addition to oil and gas, his client experience includes industries such as automotive, electronics, consumer goods, textiles and chemical companies in the United States, Europe, and Australia. Previously, Mr. Laseter worked for Siecor Corporation, a fiber optic cable manufacturer.

Mr. Laseter earned honors degrees in business from the University of Virginia's Darden School and in industrial management from the Georgia Institute of Technology. Additionally, he is certified at the Fellow level by the American Production and Inventory Control Society and has been certified as a Quality Engineer by the American Society for Quality Control.

Jay Marshall is a vice president in the energy and chemicals group of Booz·Allen, specializing in strategy development and implementation and organization design. Mr. Marshall is one of the firm's leaders in business process reengineering (BPR) and change management. For the past few years, Mr. Marshall has concentrated on significant BPR engagements at major oil and gas companies.

He received a BSE in chemical engineering from Princeton University and an MBA with a concentration in finance from the University of Texas at Austin, where he was a Sord Scholar and received the Dean's Award.

Neil C. McArthur is a senior associate in the Wassenaar (The Hague) office of Booz·Allen & Hamilton Inc. He specializes in strategy development, business process reengineering, and procurement activities for major oil and gas companies. Prior to joining Booz·Allen & Hamilton, he spent seven years with Royal Dutch Shell in a variety of international engineering and production assignments.

Mr. McArthur received an MBA with distinction from INSEAD in Fontainebleau, France, and a first-class honors BSc degree in civil engineering from the University of Glasgow, Scotland. He is a fully chartered Mechanical Engineer M.I. Mech. Eng.

Matt McKenna is a principal with Booz·Allen. His experience includes analysis and implementation in the areas of supply chain management, sourcing strategy, operations strategy, and business process redesign. He focuses on assisting clients in process-intensive industries, including oil and gas, chemicals, pharmaceuticals, and electric utilities.

Prior to joining Booz·Allen, Mr. McKenna was with American Software, Inc. (ASI), a supplier of integrated mainframe software. While at ASI, he led the planning, design, and implementation of demand forecasting, procurement, manufacturing, and distribution application software systems.

Mr. McKenna received his MBA from the Harvard Business School. He earned a BS from the School of Industrial and Systems Engineering at the Georgia Institute of Technology. He holds a Certificate of Production and Inventory Management (CPIM), and is a member of the American Production and Inventory Control Society and the Institute of Industrial Engineers.

Baxter Nairon is a Principal with Booz·Allen & Hamilton's Energy and Chemicals practice in the Houston office. He is one of Booz·Allen's leading practitioners of Business Process Redesign and organizational restructuring, having contributed to building much of the firm's methodology. He has worked with most U.S. major oil and gas companies in a variety of BPR, benchmarking, organization restructuring, cost reduction and strategy engagements.

Mr. Nairon received a BS in mechanical engineering with honors from the University of Tennessee in Knoxville, and an MBA from the University of Texas, Austin, where he received the "Outstanding Student Award" and served as president of the graduate business student body. He is a registered professional engineer.

Bruce Pasternack is a senior vice president of Booz·Allen and a member of the firm's executive committee. He has responsibility for Booz·Allen's firmwide energy, chemicals, and pharmaceuticals practice.

Prior to joining Booz·Allen, Mr. Pasternack was Associate Administrator for Policy and Program Evaluation at the Federal Energy Administration. He was responsible for energy policy development and coordination in the

Executive Branch and served as principal staff to the President's Energy Resources Council, a cabinet group assembled to coordinate energy policy.

Before coming to FEA, he served as a staff member for energy programs at the President's Council on Environmental Quality in the Executive Office of the President, worked with the General Electric Company's Space Division, and was employed at Exxon Research and Engineering.

Mr. Pasternack has engineering and operations research degrees from the Cooper Union and the University of Pennsylvania. He is a member of the Advisory Council of Stanford University's Graduate School of Business. He serves on the Governors for Energy of the World Economic Forum in Davos. He also cochaired the Wealth of Nations Conference in Scotland and served as an advisor to the Congressional Office of Technology Assessment. He is a member of the San Francisco Mayor's Fiscal Advisory Committee, and the Strategic Planning Committee of the American Gas Association.

Harry Quarls is a vice president of Booz·Allen and a senior member of its oil and gas practice. In this capacity, he has served major and independent oil and gas companies both domestically and internationally. He has experience across all segments of the business: exploration, production, refining, marketing, natural gas, and petrochemicals. Mr. Quarls specializes in strategy formulation and shareholder value analyses.

Mr. Quarls received BS and MS degrees in chemical engineering from Tulane and MIT, respectively. He was also awarded an MBA from Stanford University where he was an Arjay Miller scholar and Alexander Robichek award winner.

Joe Quoyser is a principal in the energy and chemical group of Booz·Allen, based in the firm's Dallas office. His work focuses on strategy and organization for oil and gas companies. He has recently managed a wide range of assignments for major and independent oil and gas companies in the United States and Europe. Prior to joining Booz·Allen, Mr. Quoyser worked as a petroleum engineer for Exxon.

Mr. Quoyser received a BS in chemical engineering from Rice University and a Master of Management from Northwestern University, where he was an F.C. Austin Scholar.

Matt Rogers is a principal with Booz·Allen & Hamilton. His areas of specialization include business process reengineering, supply chain management, strategic planning, organization design, and regulatory analysis for oil and gas companies and utilities. Prior to joining Booz·Allen, Mr. Rogers worked for the First Boston Corporation, a New York investment bank, where he concentrated on project financing and financial restructuring for public utilities and energy companies. He has also written extensively on supply, trading, and informations systems.

Mr. Rogers holds a masters degree in public and private management from Yale University's School of Organization and Management. He received his BA (magna cum laude) in politics from Princeton University.

Bert Shelton is a vice president in Booz·Allen's oil and gas practice, based in San Francisco. He has consulted to several major integrated oil companies, both upstream and downstream; as well as independent producers, gas transmission companies, and local utilities. Prior to joining the firm, Mr. Shelton was manager of marketing strategy development for General Electric Company's major appliance business. Previously, he held various positions with General Electric and with the Procter & Gamble Company. Mr. Shelton received an MS in history and philosophy of science from Indiana University and a BS in physics and philosophy from Thomas More College.

Richard Spitzer is a principal with Booz·Allen based in Houston. Mr. Spitzer's focus is on improving corporate and business unit effectiveness and efficiency through restructuring, process redesign, internal control systems, and organizational technologies.

Mr. Spitzer received a BS in mechanical engineering with honors from the University of Mississippi in Oxford and an MBA with honors from Southern Methodist University in Dallas. He is a registered professional engineer and certified quality engineer.

Eric Spiegel is a vice president in the energy and chemicals practice within Booz·Allen. He specializes in profit improvement, strategy, and organization development for oil and gas companies, electric and gas utilities, and other process industries. Mr. Spiegel is a leader of the firm's downstream oil and gas practice area, and a developer of the firm's business process reengineering (BPR) methodology.

Mr. Spiegel received a BA with honors in economics from Harvard University and was awarded the Harvard College scholarship for academic achievement of high distinction. Mr. Spiegel received an MBA from the Amos Tuck Graduate School of Business Administration at Dartmouth College where he was an Edward Tuck Scholar.

Earle Steinberg is a vice president with Booz·Allen based in Houston, Texas. He has extensive consulting experience in engineering, manufacturing and distribution management with emphasis on manufacturing modernization, maintenance, Total Quality Management, Just-In-Time, Design for Excellence, systems design, selection, and implementation, as well as strategic planning for engineering and manufacturing firms.

Prior to joining Booz·Allen, Mr. Steinberg worked for Coopers & Lybrand as Partner-in-Charge, Southwest Region, Manufacturing Management Consulting. He is the author of a book on service parts management.

George Thibault leads Booz·Allen's business simulation practice and is director of senior professional development for the firm. He is a retired naval officer who served at sea in destroyers and cruisers, chaired the department of strategy at the National War College in Washington D.C., and served as special assistant to the Director of Central Intelligence.

Ross Tokmakian is a principal with Booz·Allen & Hamilton's energy and chemicals practice. He has focused on improving business unit effectiveness and efficiency and on strategy development and execution for major oil and gas and resource based companies. He is a leader in the ongoing development of the firm's Capability Based Strategy methodology.

Mr. Tokmakian received a masters of science degree in industrial administration from Carnegie-Mellon University. He received BS in chemistry from Harvey Mudd College in Claremont, California.

John Elting Treat is vice president of Booz·Allen and is responsible for management consulting to the international petroleum industry. His recent work includes assignments for national and international oil companies in North and South America, Europe, the Middle East, and Asia.

Prior to joining Booz·Allen, he served as president of Regent International, a venture capital company with investments in energy and other industries and as executive publisher of *Petroleum Intelligence Weekly*. He was a partner in the investment banking firm of Bear, Stearns and Company, where he founded that company's energy group, following his tenure as president of the New York Mercantile Exchange (NYMEX). Mr. Treat has served in a variety of senior positions in the U.S. government, including the position of White House energy adviser as a member of the National Security Council Staff under Presidents Carter and Reagan and Deputy Assistant Secretary for International Affairs in the U.S. Department of Energy.

Mr. Treat received degrees in international economics from Princeton and Johns Hopkins University. He is also the author/editor of several books, including *Energy Futures: Trading Opportunities for the 1990s*, also published by PennWell.

Douwe Tideman is a principal in the energy practice of Booz·Allen's London office, and has concentrated on strategy and organization development for clients in the oil and gas and utility sectors. He has led studies for clients in the United States, France, United Kingdom, Germany and the Netherlands. Prior to joining Booz·Allen, Mr. Tideman held several positions within Shell Internationale Petroleum Maatschappij.

Mr. Tideman holds an MSc and BSc (with distinction) degree in civil engineering from the Technische Hogeschool Delft in Holland and an MBA degree from l'Institut Superieure d'Administration (INSEAD) in Fountainebleau, France.

Albert Viscio is a vice president with Booz·Allen's energy practice in San Francisco, where he specializes in strategy and organization for oil and natural resource companies.

Dr. Viscio holds a PhD in economics from New York University and a BA in economics from Holy Cross College.

PREFACE

As management consultants, we have been intimately involved in helping the world's oil and gas companies adapt to the rapidly changing demands of the marketplace during the past decade. From this work we have gained a deeper understanding of the challenges facing oil industry executives today and have developed some practical approaches that can help the senior manager who is looking for ways to make his company not only survive, but prosper in today's intensely competitive world.

That's why we wrote this book—to help those managers transform their companies into high-performance enterprises. The book is organized in five major parts:

- Section I provides a brief historical discussion of the oil industry and the forces which are driving the need for change.
- Section II describes our approach to making strategic choices. The chapters in this section cover overall strategy and then delve into the keys to success in each of the major petroleum business segments: upstream, refining, supply/trading and retail/marketing.
- Section III focuses on how to build and maintain strategic capabilities. In other words, how to run an oil company. As most senior managers have learned the hard way, the best strategy is useless without the capability to execute efficiently and effectively. The individual chapters here examine the design of critical business processes and how the oil company should be organized to facilitate the processes. We also will be discussing the planning process, how to optimize the value chain from procurement to maintenance, how to achieve high quality at low cost, and how to develop good people. And finally, we will see how companies today are increasingly linked in a web of relationships, joint ventures, and alliances that constitute an extended enterprise and what this means to the individual companies involved.
- Section IV examines the special case of national companies and looks at their options, which range from simply becoming more commercially focused to outright privatization.

- Section V considers the value of integration and describes what the high-performance oil company of the twenty-first century should look like.

A list of acronyms is included at the front of the book to help the reader decode the language of consulting.

ACKNOWLEDGMENTS

This book has been an intensely collaborative effort, not only by the partners and principals of Booz·Allen whose names appear on the cover , but also our colleagues throughout Booz·Allen who have contributed in ways great and small to our understanding of this fascinating industry.

While space will not let us name them all, we would like to acknowledge the contributions of:

Allen Chan
Kathy Bojack
Paul Denton
Mark Gallion
Pam Kreiter
Gary Neilson
Frank Mizumo
Benton Routh

We would also like to thank our editors, Sue Rhodes Sesso and Madeliene Reardon at PennWell Books, as well as Paul B. Brown, for their patience, understanding, and help.

And finally we want to thank our clients, who have worked with us to improve their businesses. We measure our success by theirs. Our thanks, therefore, go to the hundreds of dedicated people at Alyeska, Amoco, Arco, Astra, British Gas, British Petroleum, Cabot, Chevron, Chinese Petroleum Company, Conoco, Elf, ENI, Exxon, Gas del Estado, Mobil, Murphy, Nicor, PDVSA, Pemex, Petrobras, Petroperu, Phillips, PTT, QGPC, Sun, Star, Shell, Texaco, Unocal, and YPFB who have been our partners in this journey of discovery.

ACRONYM LIST

ACRONYM	DEFINITION
4M+S	Alliance of Mitsui, Mitsubishi, McDermott, Marathon and Shell
ABB	Asea Brown Boveri
ADNOC	Abu Dhabi National Oil Company
AGIP	Azienze Generale Italianan Petroli (Italy)
Ancap	Administracion Nacional de Combustibles Alcohol y Portland (Uruguay)
API	American Petroleum Institute
Banagas	Bahrain National Gas Co.
Banoco	Bahrain National Oil Co
Bapco	Bahrain Petroleum Co.
BCF	Billion Cubic Feet
BG	British Gas
BHP	Broken Hill Proprietary, Ltd.
BNOC	British National Oil Corporation
BNP	Banque Nationale de Paris
BOC	Barbados Oil Co.
BOPs	Blowout Preventers
BP	British Petroleum
BPR	Business Process Redesign
BPX	British Petroleum Exploration Ltd.
CEGB	Central Electricity Governing Board
CEPE	Corp Estatal Petrola Ecuatoriana (Ecuador)
CNOOC	China National Offshore Oil Corp
CNPC	China National Petroleum Corp
CPC	Chinese Petroleum Corp
CRINE	Cost Reduction in a New Era
CVP	Original Venezuelan National Oil Company
DONG	Dansk Oilie og Naturgas A/S (Denmark)
DSM	Dutch State Mines
DTW	Dealer Tank Wagon
Ecopetrol	Empresa Colombiana de Petroleos (Colombia)
E&P	Exploration and Production
EdF	Electricité de France
EGPC	Egyptian General Petroleum Company
EGPC	Emirates General Petroleum Corp (Dubai)

ACRONYM	DEFINITION
Elf	Ste Nationale Elf Aquitaine (France)
Enagas	Empresa Nacional del Gas SA (Spain)
ENAP	Empresa Nacional del Petroleo (Chile)
ENC	Empresa Nacional de Combustivos (Cape Verde Islands)
ENH	Empresa Nacaional de Hidrocarbonetos de Mozambique
ENI	Ente Nazionale Idrocarburi (Italy)
EPC	Ethiopian Petroleum Corp
ETAP	Enterprise Tunisienne d'Activites Petroliere (Tunisia)
FCCU	Fluid Catalytic Cracking Unit
FERC	Federal Energy Regulatory Commission
FOGCO	Federal Oil and Gas Corporation
FTE	Full-Time Equivalent
Gasunie	NV Nederlandse Gasunie (Holland)
GdF	Gaz de France
Gepsa	Empresa Guineano-Espanola de Petroleos (Equatorial Guines)
HAZOPs	Hazarous Operations
HRD	Human Resource Development
HSE	Health, Safety, and Environment
Hydro Congo	Ste Nationale De Recherche et d'Exploitation Petrolieres (Congo)
INA	Industrija Nafte (Croatia)
INH	Insituto Nacional de Hidrocarburos (Spain)
INOC	Iraq National Oil Company
INPC	Irish National Petroleum Corp Ltd.
IQS	International Quality Study
IRR	Internal Rate of Return
IT	Information Technology
JNOC	Japan National Oil Corporation
KNPC	Kuwait National Petroleum Company
KOPC	Kuwait Oil Co
KPC	Kuwait Petroleum Corp
KSF	Key Success Factor
LNOC	Libyan National Oil Company
MAR	Hellenic Aspropyrgos Refinery SA (Greece)
MBNQA	Malcolm Baldrige National Quality Award
mcf	Thousand Cubic Feet
MMC	Monopoly and Mergers Commission
MTBE	Methyl Tertiary Butyl Ether
MWD	Metering While Drilling

ACRONYM	DEFINITION
NAM	Exxon/Shell Joint Venture in Dutch North Sea
NBU	Natural Business Unit
NGC	National Gas Company (Trinidad)
NOC	National Oil Company
NPN	National Petroleum News
NPV	Net Present Value
NRA	National Resources Authority (Jordan)
NUMMI	Car company, not further identified
OEM	Original Equipment Manufacturer
OGDC	Oil and Gas Development Commission (Pakistan)
OMV	Osterreichische Mineralol Verband (Austria)
ONAREP	Office National de Reserches Petroliers(Morocco)
ONGC	Oil and Natural Gas Commission (India)
OSHA	Occupational Safety and Health Administration
Pemex	Petroleos Mexicanos (Mexico)
Petrobangla	Bangladesh Oil, Gas & Minerals Co.
Petrobas	Petroleo Brasileiro SA (Braxil)
Petrocorp	Petroleum Corp of New Zealand Ltd
Petrogab	Petroleum Corporation of Gabon
Petrogal	Petroleos de Portugal EP
Petroguin	Empresa Naciohanl de Pesqulsas e Exploracao Petroliferos (Guinea-Bissau)
Petromin	General Petroleum and Mineral Organization of Saudi Arabia
Petromoc	Empresa Nacional Petroleos de Mozambique
Petronas	Petroliam Nasional Bhd (Malaysia)
Petronic	Empresa Nicaraguense del Petroleo (Nicaraqua)
Petroperu	Empresa de Petroleos del Peru
Petrosen	Ste Nationale des Petroles de Senegal
Petrotrin	Trinidad and Tobago Petroleum Co. Ltd.
Petrozaire	Petroleum Company of Zaire
P&L	Profit and Loss
PCA	Company name, not further identified
PDVSA	Petroleos de Venezuela, S.A.
PBMS	Performance Based Management Systems
PMPA	Petroleum Marketing Practices Act
PNCO	Philippine National Oil Co.
POGC	Polish Oil & Gas Corporation
PSCO	Pakistan State Oil Co Ltd.
PT	Partnering Team

ACRONYM	DEFINITION
PTT	Petroleum Authority of Thailand
PUC	Public Utility Commission
QA	Quality Assurance
QC	Quality Control
QGPC	Qatar General Petroleum Corp
Recope	Refinadora Costarricense de Petroleo SA (Costa Rica)
RVP	Reid Vapor Pressure
Saskoil	Saskatchewan Oil and Gas Corp
Sinopec	China Petrochemical Corp
S&P	Standard & Poor's
SNAM	Italian National Gas Company
SNH	Ste Nationale des Hydrocarbures (Cameroon)
Socar	State Oil Company of Azerbaijan Republic
SOE	State-Owned Enterprise
Sonagol	State Oil Company of Angola
Sonatrach	Entreprise National Sonatrach (Algeria)
Sonacop	Ste. Nationale de Commercialisation des Produits petroliers
Sonidep	Ste Nigerienne de Produits Petroliers (niger)
Soquip	Ste Quebecoise d'Initiatives Petroliers (Canada)
SP Exploration	Svenska Petroleum Exploration AB (Sweden)
SPC	Singapore Petroleum Company
Staatsolie	Staats Olie Mij Suriname NV (Surinam)
Statoil	Den Norsk Stats Oljeselskap AS (Norway)
Total	Cie Francaise des Petroles (France)
TPAO	Turkish Petroleum Corp
TQM	Total Quality Management
Veba	Vereinigte Electrizitats- und Bergwerke - AG (Germany)
YPF	Yacimentos Petroliferos Fiscales (Argentina)
YPFB	Yacimentos Petroliferos Fiscales Boliviana (Bolivia)

SECTION I
INTRODUCTION

CHAPTER 1

THE HISTORICAL STRUCTURE OF THE OIL INDUSTRY

World oil markets and the companies which are the primary actors in those markets are undergoing an unprecedented transformation. As a point of departure, it is useful to first review briefly the history of the oil industry. Its history helps explain what is driving the changes executives are wrestling with today.

A BRIEF HISTORY OF THE OIL INDUSTRY

he 20th century began with the creation of the first major government-owned oil company and is ending in a wave of privatizations. To many observers, national oil companies (NOCs) seemed to be a creation of the post-embargo world of the 1970s. In fact, as shown on Figure 1–1, they have been around far longer.

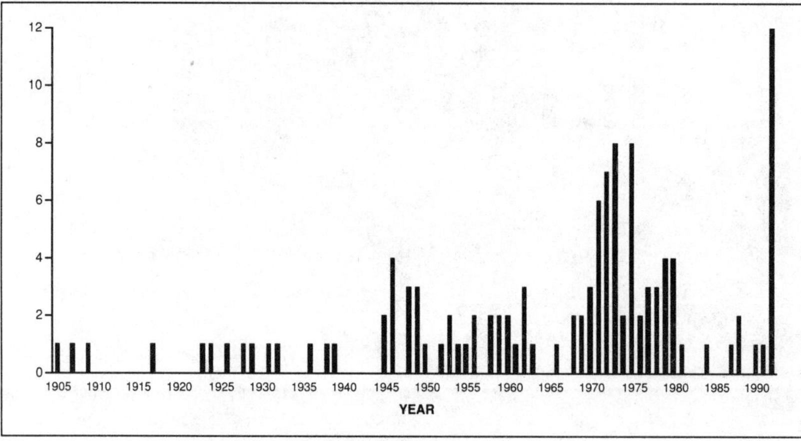

Fig. 1–1 Creation of National Oil Companies.

HOW NOCs DEVELOPED IN THE FIRST HALF OF THE TWENTIETH CENTURY

The development of national oil companies went through four distinct states during the first half of this century.

- Pre-World War I. The conversion of the British Navy from coal to oil prompted the creation of the first major national oil company, British Petroleum (BP). BP began its life as Anglo-Iranian, a private company.

- Russian Oil Turns Red. The Russian Revolution of 1917 triggered creation of the second major national oil company, the Soviet Oil Ministry and its affiliated production, refining and distribution associations, as the Bolshevik government nationalized the private oil companies operating within their territory.

- The First Postcolonial Wave: The history of national oil companies in the producing areas of the developing world ironically begins in Argentina—one of the leaders of today's privatization movement. Yacimentos Petroliferos Fiscales (YPF) was founded in 1922.

  Initially, it operated in competition with private companies such as Shell, Exxon (then called Standard Oil of New Jersey) and Astra. However, in 1935, YPF was given exclusive rights to new exploration and development in Argentina. Until the establishment of Petroleos Mexicanos (Pemex) in March 1937, YPF remained the solitary example of a national oil company in the developing world for 15 years.

  (It should also be noted that several European national companies, including Total [1920], Agip [1926] and Veba [1929], were created during this period.)

- World War II, a war fought in part about oil and access to it, triggered two very different responses. The Axis powers of Germany, Austria and Japan, plus sympathizer Spain, all created national oil monopolies. In contrast, the Allies tended to rely more on the national mobilization of private company resources without changing actual ownership.

It was also during the first half of this century that the management structure of oil companies evolved to reflect their growing complexity and increasing geographic scope. During this time, oil companies moved from highly centralized organizations to companies with multidivisional structures.

THE DEVELOPMENT OF NATIONAL OIL COMPANIES IN THE SECOND HALF OF THE TWENTIETH CENTURY

Creation of national oil companies during the second half of this century can be divided into five distinct eras:

■ ***European Recovery.*** In the aftermath of World War II, ENI was created in 1953 by the charismatic, visionary, and controversial Enrico Mattei, who built a major international oil company on the domestic Italian base of Agip.

■ ***Catching the Second Postcolonial Wave.*** Following the creation of ENI, a second round of nationalism was touched off in earnest. Petrobras was established in Mexico in 1953 and Sasol in South Africa in 1955.

EGPC was created in Egypt in 1957, following Nasser's revolution. ONGC was formed in India in the same year. Venezuela created CVP in 1960. OGDC was formed in Pakistan in 1961, and Algeria created Sonatrach in 1963. Iraq's INOC was formed in 1964, and Petroperu was created in Peru in 1968.

■ ***The New World Economic Order.*** The Arab oil embargo is inevitably linked in peoples' minds with the nationalizations that followed, but actually the trend started before that. It was, in fact, Libya's Colonel Khaddafi who set off the third wave of nationalizations in the second half of the twentieth century.

After his seizure of power from King Idris in 1969, the controversial colonel set his mind to redistributing the growing revenues from oil production in Libya. He did that by creating Libya's National Oil Company in 1970. Prior to the 1973 embargo, a diverse group of countries nationalized as well. Trinidad did in 1970, Nigeria and Abu Dhabi in 1971, Norway and Saudi Arabia in 1972, and Iran the following year.

■ ***The Postembargo National Oil Company Explosion.*** OPEC's success in raising and sustaining oil prices in the aftermath of the 1973 embargo led to the creation of the NOCs in over a dozen countries including the Philippines (1973), Qatar (also 1973), Australia in 1974 (although they changed their mind in 1975), Malaysia (1974), Canada (1976), Ecuador (1976), France (Elf in 1976), Kuwait (1976), New Zealand (1978), and Gabon (1979).

In 1974 there was even a short-lived attempt to create a U.S. national oil company, but it was handily defeated in Congress, in part due to an astute decision by one opponent to call the proposed project "FOGCO," the Federal Oil and Gas Corporation. The breakup of the Soviet Union spawned the creation of a dozen NOCs as the newly independent former Soviet Union republics rushed to create their own oil companies.

■ ***Oil Stocks of the Third Kind.*** Throughout most of the past 100 years, the trend was toward ever greater government ownership in the oil industry. However, in the century's final decades, a powerful countercurrent has developed. Once again, the British led the way. It was Britain that created the first major government-owned oil company, and it is Britain that is leading the privatization charge by abolishing the British National Oil Corporation and returning BP and Britoil to private ownership.

The move began in the early 1980s. The British Treasury needed money and Prime Minister Margaret Thatcher wanted to take a bold step toward bolstering the role of the private sector in the British economy.

Few anticipated the momentous impact her proposals would soon have on the structure of the world's oil industry. Soon thereafter, New Zealand privatized Petrocorp, and the dynamic Carlos Menem of Argentina announced the privatization of both YPF and Gas Del Estado.

Within the space of half dozen years, the nationalization trend quickly accelerated and spread, involving in varying degrees: Repsol, Total, Elf, Singapore Petroleum, OMV, Petrocanada, and more recently, Petroperu and ENI.

And an even larger number of companies were "commercialized," transformed from government bureaucracies into more market-oriented companies. Examples include: Pemex, Saudi Aramco, KPC, PTT, Petronas, CPC, and many of the oil companies of eastern Europe and the former Soviet Union. In essence, they are taking a middle course between being a traditional national oil company and one that is privately owned.

Still, despite that growing trend, national oil companies now dominate the world's petroleum reserves, and play a growing role in the downstream market as well (see Figs. 1–2 and 1–3).

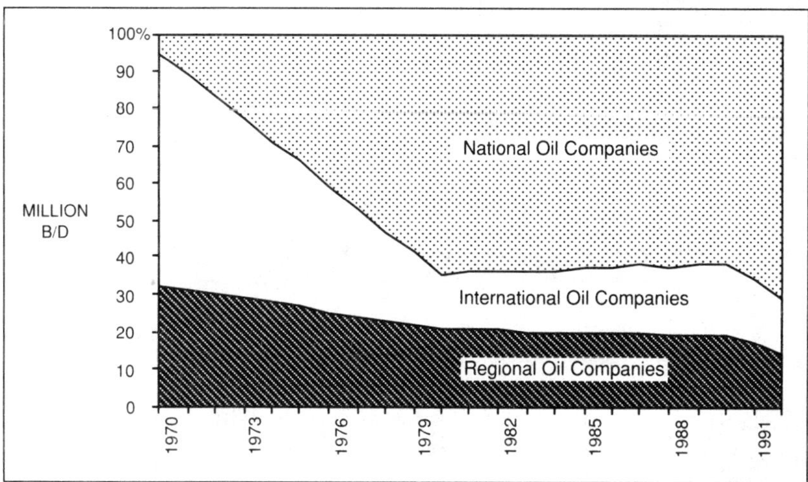

Fig. 1–2 Historic Oil Production 1970-92. Source: PetroCompanies, *Oil & Gas Journal*, BAH Analysis.

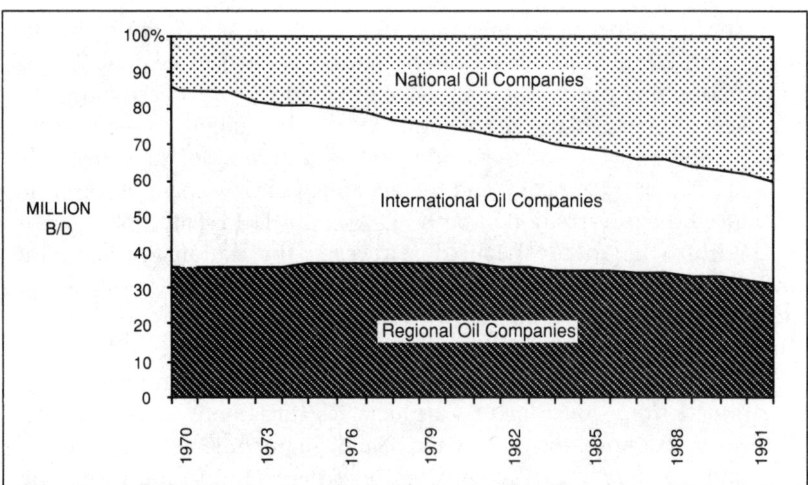

Fig. 1-3 Historic Refining Capacity 1970-92. Source: PetroCompanies, *Oil & Gas Journal*, BAH Analysis.

Yet too often national oil companies are thought to be indistinguishable from their governments. That impression is not surprising since the national government is their primary and often exclusive shareholder. But today NOCs from Venezuela to Taiwan are asserting their own identifies and are striving to become full-fledged partners with the independent oil companies of the world.

Obviously, this fact will be an important part of our discussion. After all, this book is about the process of transformation which oil companies—whatever their ownership—must undertake in today's changing petroleum markets.

THE DRIVERS OF CHANGE

A fundamental transformation of oil markets is underway. This change affects all companies, large and small; regional, national, and international. It is driven by the convergence of nine forces:

■ ***Intensification Of Global Competition.*** The entry of new players and the breakdown of market barriers has intensified global competition in petroleum markets worldwide.

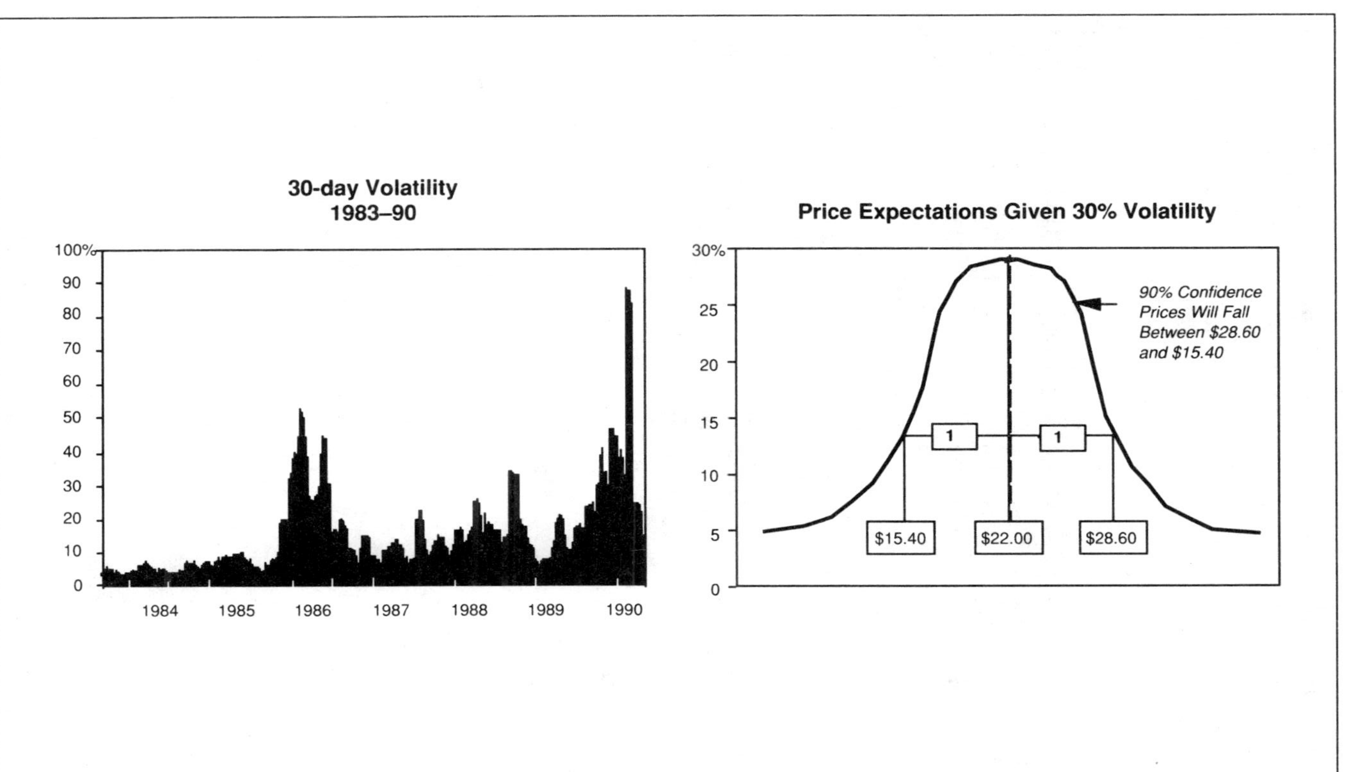

Fig. 1–4 Price Volatility/Greater Risk.

■ ***Volatility Of Oil Prices.*** The growing competition, coupled with the decline in OPEC's ability to control prices, have led to an unprecedented volatility of oil prices (see Fig. 1–4).

■ ***Deintegration.*** The decoupling of upstream resources from downstream outlets, and the subsequent robust growth of both spot and derivative markets, has further deintegrated the oil industry. In spite of the well-publicized downstream moves by several large national oil companies, such as PDVSA's purchase of CITGO and its subsequent absorption of Champlin, KPC's expansion in Europe, and Saudi Arabia's joint venture with Texaco, the industry remains far less integrated today than it was two decades ago.

■ ***New Technology.*** Throughout history, new technology has lowered commodity production costs and expanded our geographic and geologic horizons. Oil is no exception. Recent innovations which are driving down the costs of exploration and production include: 3–D seismic, horizontal drilling, coiled tube drilling and tension leg platforms.

■ ***Communications revolution.*** The global telecommunications revolution has dramatically reduced the cost of information processing, making decentralization far more feasible.

■ ***Political transformation and the breakdown of national frontiers.*** Recent profound political transformations have opened many previously closed economies to foreign investment and private ownership (see Fig. 1–5).

■ ***Growing Global Environmental Awareness.*** This factor has also had a profound impact on the oil industry. Restricted access to resources, partic-

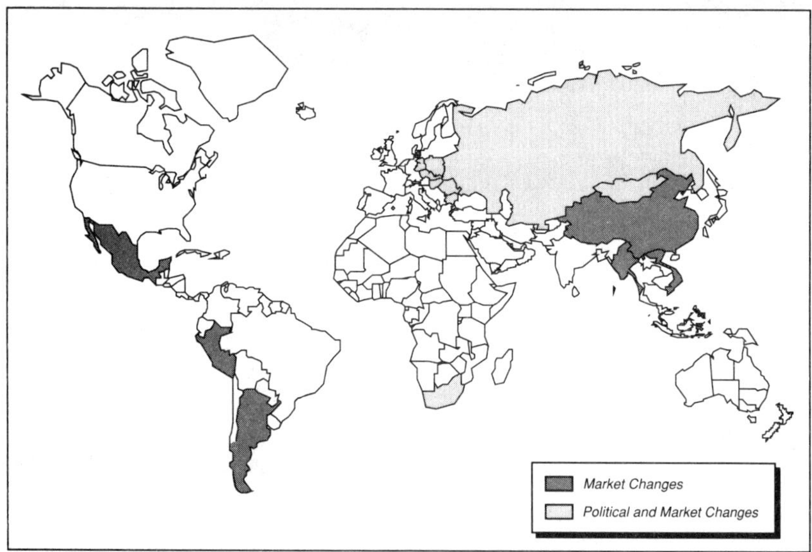

Fig. 1–5 Political and Market Changes, 1989-92

ularly in North America, has accelerated the globalization efforts of North American oil companies. Increased costs of exploration, production, refining, transportation and marketing—especially in North America and Europe—has required major changes in the formulation of key petroleum products. And increased regulation worldwide has forced companies to spend more on compliance. In response to these forces, oil companies are fundamentally reevaluating their businesses.

■ ***Ownership And Governance.*** As previously mentioned, the entire question of ownership and governance is being revisited. In stark contrast

OBJECTIVES	CURRENT STATUS			
	Commercialization	Demonopolization	Partial Privatization	Complete Privatization
Complete Privitization			PetroCanada Repsol Total Elf YPF ENI	British Gas BP
Partial Privatization		Rosneft Petronas ONGC	Gazprom Norsk Hydro Yukos Lukoil Slavneft Petroleas Del Peru Siberian Far Eastern Surgutneftegaz	
Demonopolization	INOC Petrobras	Sonatrach EGPC NNPC Libya NOC PDVSA PDO ADNOC ONGC Pertamina QGPC SINOPEC NIOC Statoil		
Commercialization	Petrobras Saudi Aramco KPC PEMEX			

Fig. 1–6 Selected Oil Companies: Privatization Status.

Type	Rank	Company	Type	Rank	Company
LNOC	1	Saudi Aramco	LNOC	26	INOC
LNOC	2	Rosneft	LNOC	27	Petronas
NOC	3	RD Shell	IOC	28	Siberian Far East
LNOC	4	NIOC	IOC	29	Surgutneftegaz
LNOC	5	PVDSA	IOC	30	Yukos
NOC	6	Exxon	LNOC	31	Libya NOC
LNOC	7	Pemex	ROC	32	Conoco
NOC	8	Mobil	IOC	33	Statoil
NOC	9	BP	ROC	34	Phillips
LNOC	10	Sinopec	ROC	35	Marathon
NOC	11	Chevron	IOC	36	ONGC
LNOC	12	Pertamina	ROC	37	YPF
ROC	13	Amoco	ROC	38	Petrofina
LNOC	14	NNPC	ROC	39	Unocal
LNOC	15	Sonatrach	IOC	40	PDO (State)
IOC	16	ENI	IOC	41	QGPC
NOC	17	Texaco	IOC	42	Repsol
LNOC	18	ADNOC	ROC	43	Ameranda Hess
LNOC	19	KPC	IOC	44	Ecopetrol
IOC	20	Petrobas	IOC	45	EGPC
IOC	21	Elf Aquitaine	IOC	46	Slavneft
LNOC	22	Gazprom	IOC	47	Petroleos del Peru
IOC	23	Total	ROC	48	BHP
LNOC	24	Lukoil	IOC	49	Petro Canada
ROC	25	Arco	ROC	50	British Gas

Large National Oil Companies (LNOC)
National Companies (NOC)
International Oil Companies (IOC)
Regional Oil Companies (ROC)

Fig. 1–7 The 50 Largest Oil Companies in the World. Source: Annual Reports, PetroCompanies, PIW, First Boston, *Moody's Oil & Gas Journal*, Eastern Bloc Energy, Petroleum Economist, BA&H analysis.

to what happened at the beginning of the century, there is now a marked reversal of the trend towards greater government ownership (see Fig. 1–6).

However, ownership of the upstream remains overwhelmingly under government control. Of the 50 largest petroleum companies in the world, 31 are state owned (see Fig. 1–7).

And, in spite of the downstream moves by several large national companies discussed earlier, the industry remains less integrated today that two decades ago (see Fig. 1–8).

■ ***Deregulation.*** Oil and gas have either enjoyed (or suffered, depending on your point of view) regulation for over 50 years. The regulations often took the form of price controls and restrictions on competition. They also limited interfuel competition, for example, by limiting the ability of oil companies from entering the power generation business. In the past decade, governments throughout the world have begun eliminating those barriers. In recent years, oil and gas companies such as Enron have entered the power generation business. And throughout the world, price controls and import restrictions are being eliminated or at least relaxed.

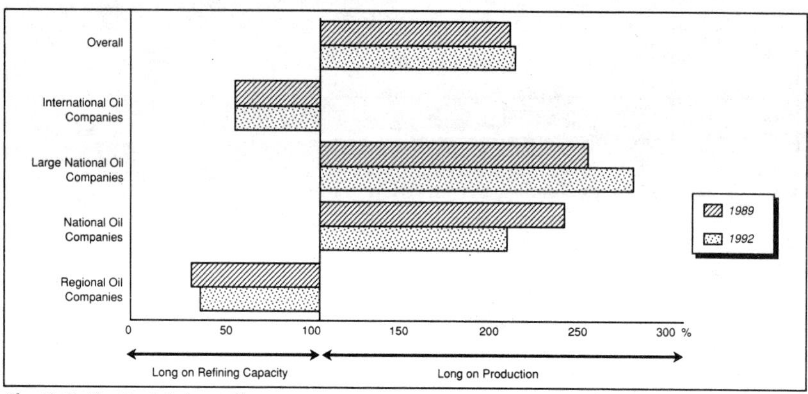

Fig. 1–8 Vertical Integration.

NEW STRATEGIES FOR NEW TIMES

ompanies are adapting their corporate strategies to the realities of the new marketplace. Their primary goals are improved profitability and the achievement of sustainable competitive advantage.

Some of key characteristics of this transformation include:

● Greater concern with creating shareholder value, rather than growth in physical terms. In the case of state-owned companies, the increasing emphasis is on sustaining income for government shareholders.

● Recognizing the importance of segment profitability, because deep spot markets now permit the decoupling of the upstream from the downstream. Some companies are even separating exploration from production and refining from marketing

● Cutting both capital and operating costs, although the results of this effort are only beginning to show up in annual reports (see Fig. 1–9).

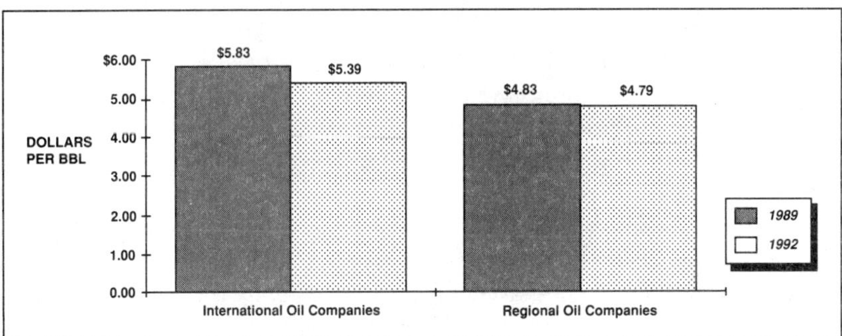

Fig. 1–9 Production Costs. Source: J.S. Herold.

● Divesting unprofitable and nonessential assets. Companies have left such noncore businesses as minerals, office products, and hotels. They are also selling or swapping poorly performing oil or gas assets.

● Identifying and building core capabilities, while outsourcing noncore functions (drilling, payroll processing, land development, etc.)

● Reallocating investments towards more profitable sectors. For example, major U.S. oil companies are shifting their exploration and production funds away from the United States and toward overseas prospects.

- Reducing the scope of operations. Companies are concentrating on more profitable geographies or niches.

- Redesigning their business processes and organizational structure to reduce costs and increase flexibility. Examples abound from the radical restructuring of companies such as BP and Pemex to the evolutionary changes in companies such as Amoco and Chevron.

- Increasing productivity improvements from technology investments. The falling cost of communications has made decentralization far more feasible. Lower computing costs and the growing client/server environment, have helped enormously as well. The ability to integrate data from a variety of sources has been dramatically improved by the use of 3-D seismic techniques. In addition, improved financial systems can now give business unit managers the information they need to run their businesses more profitably because they provide those managers with a better understanding of their costs.

In all these approaches, the key to success is recognizing that strategy should drive the design of key business processes, and that the company's organizational structure should facilitate—not impede—those processes as illustrated in Booz·Allen's Performance Wheel (see Fig. 1–10).

While the realignment of oil company structures that has occurred over the past five years is new to the industry, other businesses have been restructuring for decades. The fact that oil companies have lagged behind other industries in restructuring is nothing new. For nearly a century, oil companies have been a "lagging indicator" of what is going on in the rest of business.

The reasons for this are both cultural and technical. The engineering and technical backgrounds of oil company executives and the perceived value of tight vertical integration led most oil companies to initially create organizational structures that were hierarchical and highly functional. As a result, the oil industry was among the last of the major industrial sectors to move to a multidivisional structure during the interwar period.

After World War II, oil companies restructured again to accommodate the growth of their allied chemical businesses, and the growing geographic reach of their operations. However, most companies (Shell was an exception) retained a high degree of central control until recently.

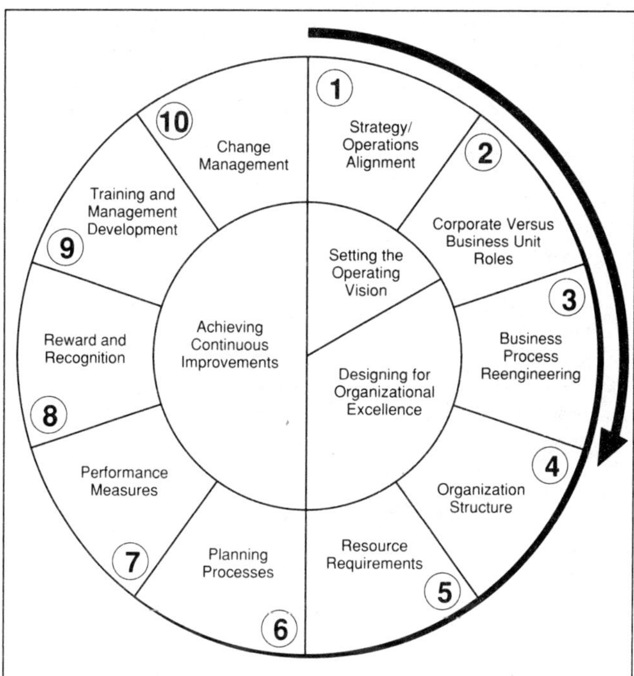

Fig. 1–10 The Booz·Allen Performance Wheel.

The last five years have seen an explosion or organizational restructurings. The most noteworthy change has been the move toward business units, breaking the old divisional structures down to a lower level, and substituting financial oversight for direct operational control.

The transformation process continues, but there are two big obstacles to change. First is the failure of companies to properly adjust business processes to the new organizational structure. The second hurdle? The entrenched resistance of corporate cultures, particularly at the middle management level, where executives are not yet comfortable with the new order.

As we have wrestled with these challenges for our clients, we have gained some insights into how to think about the oil business in some fundamentally different ways. That will be the topic of our discussion in the subsequent chapters.

SECTION II

VALUE CREATION OPTIONS: MAKING STRATEGIC CHOICES

CHAPTER 2

CAPABILITY-BASED STRATEGY

Petroleum companies have traditionally pursued position-based strategies. The key to success was thought to be the establishment of a strong position in an interesting geologic area or a large consuming market. And even today, such an approach is serving some companies well.

However, a number of converging trends are posing an ever greater threat to that time-honored approach. For example, in the upstream:

- Rapid evolution of technology such as 3D seismic, horizontal drilling, subsea completions, coiled tubing and slimhole drilling, remote sensing, automated metering, credit card readers, and the integration of new technologies and systems [3D + Measurement While Drilling (MWD) + Horizontal Drilling].

- Broad dissemination of new information that makes sustainable technological advantage difficult to maintain or defend.

- Breakdown of traditional political frontiers. The breakup of the Soviet empire is just the most obvious example.

- Evolution of new, nontraditional competitors such as the growing number of national oil companies worldwide.

These developments argue for a new approach to corporate strategy for petroleum companies who want to prosper in the twenty-first century. The new strategy must focus on building market-driving capabilities.

WHY A NEW APPROACH IS NEEDED

Differences in performance across oil companies are striking (see Fig. 2–1). The firms that have done extremely well have added to reserves at a low cost and have demonstrated an ability to create shareholder value in adverse market conditions. They are well poised for the future. Poor performers, on the other hand, have a way of spending much of what they earn looking for high cost reserves.

What separates the winners from the losers? Again focusing on the upstream, let's examine a number of possible differentiators:

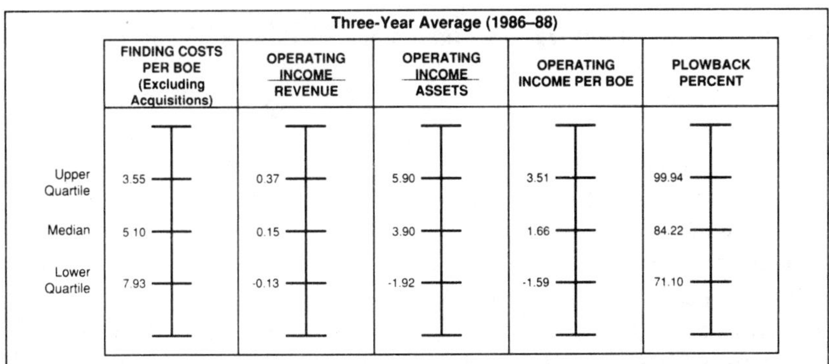

Fig. 2-1 Gas Production Peers Financial Statistics. Source: Newport Associates Database, BA&H analysis.

- ***Skills.*** Do some companies have access to superior skills in finding, developing, and producing reserves?
- ***Technology.*** Are there proprietary technologies which make some firms more capable of finding reserves?
- ***Asset base.*** Is there unfair advantage to some firms which allow them access to better areas than their competitors?

Let's examine them one at a time.

Does skill in finding reserves make the differences? Although there are skill differences among people, most companies are staffed with competent people, and truth to tell, there are not many differences between them. After all, the techniques and tools of the trade are widely known. They are taught in the same schools from which the talent pool is drawn. And, most importantly, there is a free market for labor; people move between companies. All companies have equal access to the required skills. (Although some companies may be poor judges of talent, not unlike professional sports teams.)

Is access to technology so restricted that only a few can have the latest and greatest? In the upstream oil business, technology is disseminated quickly. A Booz·Allen & Hamilton study on upstream technology showed that technology and service firms quickly make technology available across the industry. For example, techniques such as 3D seismic, slimhole drilling, and horizontal drilling are available for the purchase price (service fees, hiring an expert, or trial and error). No one company has been able to maintain a clear lead due to being the sole possessor of a special technology, although some companies are much better than others at applying the right technology and the right time (as we will discuss later). And some companies are faster than others at adopting the technologies and having a temporary lead over others, gaining enough of a window to capitalize on opportunities before others.

Do good performers have preferred access to better drilling opportunities? Clearly not. Most opportunities are competitive. Frequently they are joint ventures, so that several players have access to any given opportunity. And because of the opportunity to form partnerships, project size is often not an insurmountable barrier for most independents. Furthermore, unique access to opportunities in emerging petroleum-producing countries, while a valuable characteristic prior to the oil embargo, has since disappeared as countries have become quite sophisticated in developing exploration concession terms structured to attract, but not to overcompensate oil exploration companies.

Luck not withstanding, good performers have built broad-based capabilities which encompass more than being good in just one dimension. These companies are good because they can combine performance in ways that others cannot. Take for example the case of Oryx and horizontal drilling. The reason they were successful in the Austin Chalk was not because they had access to proprietary technology. Horizontal drilling was available to the industry. It was not because they targeted the right area; others were drilling without success. And it was probably not because their geoscientists were smarter than others.

Oryx was successful because it understood what horizontal drilling could do for it in the Chalk and acted on it immediately. Acting immediately meant totally changing the drilling program in a very short period of time to target an area of opportunity. Oryx combined its geoscientific skills and understanding of the potential for horizontal drilling in a given area with a business process (planning and budgeting for drilling) to create substantial value. They did this while many other companies waited for the next budget cycle.

An example of a broader application of capabilities can be seen in Cabot Oil and Gas' (COGC) approach to the Appalachian basin. Cabot is a midsized independent (about 300 BCF reserves) that focuses on low-risk, tight gas exploration and development.

First, some background on the Appalachia gas supply basin. This is by far the most mature of the U.S. producing areas, supplying only about 5% of U.S. gas production. Majors pulled out of Appalachia decades ago for the same reasons they are pulling out of domestic E&P today—low reserves per well, limited upside potential, and high cost of operating thousands of low-rate wells. Reserve development in Appalachia is a low-risk, low-return investment. The major challenge is gaining enough production scale to build gathering systems and get the gas to market. Spot gas prices at the wellhead in Appalachia are 10¢–30¢/MCF higher than Gulf Coast prices, due to the transportation differentials.

But while other natural gas producers were on the verge of bankruptcy during the late 1980s, here's what COGC did:

- Eliminated costs of 30¢–50¢ per MCF
- Grew reserves growth of 10%/year for five years
- Increased operating margin by about 40¢/MCF during the late 1980s.

How do they do it? Basically, by pulling together an integrated combination of operating capabilities and assets in Appalachia that included development drilling, low-cost operations, gas gathering, and gas marketing—The value in this approach stems form the fact that this integrated position is far more difficult to duplicate by a competitor than a single skill or asset-based position. Specifically, COGC has been successful in Appalachia because it concentrated directly on enhancing the bottom line through a number of measures, not solely on reserves replacement or sales volumes. More specifically, they:

1. Reduced costs by scale/focus

   - Focused on Appalachian gas—over 70% of reserves are in West Virginia and Pennsylvania
   - Became a scale player—one of the largest in the basin
   - Brought 50 years experience in Appalachia geology and operations to the project (Cabot was one of the first players in Appalachia)
   - Had a huge inventory of low-risk, undrilled prospects

2. Employed an operations cost control approach

   - Had an operational focus—COGC thinks of itself as a "gas manufacturing company"
   - Developed new operating practices and production technologies to increase recoveries from shallow low-rate wells (e.g., the plunger lift system to keep gas wells on line)
   - Established a large industrial customer base served directly from the gathering system

3. Invested for revenue enhancement

- Created a gas gathering network across Appalachia to transport gas to industrial markets and to interstate pipelines
- Created a gas marketing strategy built around "security" of local gas supplies (gas from the Gulf Coast is subject to curtailment)
- Had a large number of pipeline interconnects which enabled COGC to arbitrage prices between different interstate pipelines
- Had large storage capacity which allowed COGC to capitalize on summer-winter price swings

Competition in the exploration and production business occurs primarily in getting access to value creating opportunities. (The same holds true for enhanced recovery.) Beyond that, the business is mainly monetizing what is in the ground. We have found that successful E&P companies build a wide range of capabilities. No two companies have to follow an identical path. But good performers build capabilities which allow them to deploy their strengths quickly. The range of capabilities include:

- Drilling programs responsive to new information
- Balancing risk and growth opportunities through effective farm-in/farm-out programs
- Matching cost structures to the scale of the business
- Gaining regional scale while also diversifying risk

We view capabilities as the responsibility of management. Managers must ensure their companies go beyond being merely good in each functional area. When business processes are not effectively linked across a company, technical strengths get wasted. The types of failures which can cause underperformance are broad:

- Investing in high-risk/low-return programs (at times at the expense of more attractive programs)
- Abandoning programs too soon (usually by "shifting strategies") to avoid accountability
- Being subscale everywhere

The explanation of why a capabilities focus succeeds is twofold. First,

FOOTPRINT	PORTFOLIO
• Location, size, and character of facilities – Supply – Processing – Distribution • Market coverage and representation	• Products/programs • Proprietary technologies • Brand(s) franchise

Fig. 2–2 Positional Assets.

positional assets and skills are not sufficient; they are easily replicated (see Fig. 2–2). The oil business does not allow any sustaining advantage in these areas because the resources behind them are available to all.

Second, the nature of competition is changing (see Fig. 2–3). Initially technological performance was critical: achieve it and you could make money. Now, particularly with natural gas in surplus, marketing and meeting customer requirements further downstream is key to maximizing value. Thus market responsiveness becomes the basis of competition. Position-focused competition is based on cost optimization. The portfolio and cost restructuring of the E&P business helped performance, but many firms have gone through this, and among those companies it is no longer a competitive edge.

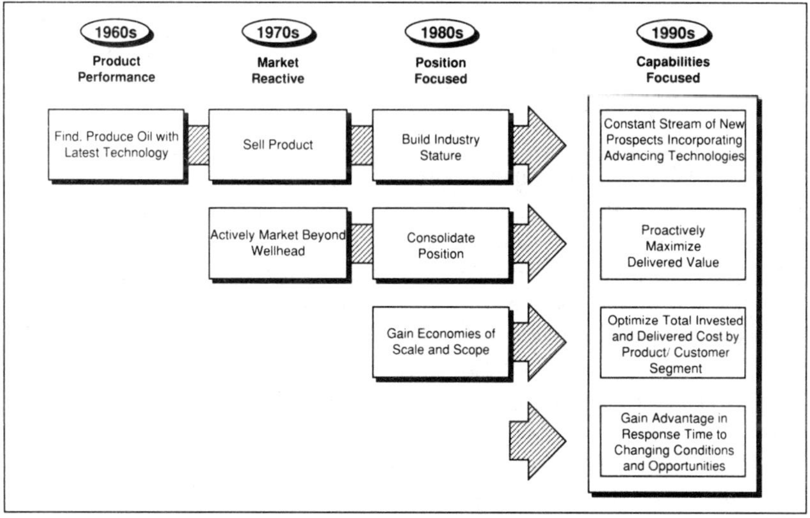

Fig. 2–3 Evolution of Competition

Technology performance, market reactiveness, and position focus, have all become critical success factors. But while they are necessary, they are not sufficient for success. Capabilities focused competition adds a new dimension of doing things well and doing them quickly. For the most part, the E&P business requires making decisions quickly. Executives must be able to evaluate performance (of assets, people, and strategies) in a timely fashion so that changes can be made before it is too late.

HOW TO ASSESS, TARGET, AND BUILD CAPABILITIES

ustainable competitive advantage, which results from market-driving capabilities, spans three dimensions:

TIME	A business' ability to create or get to demand first
VALUE	The created demand (valued-added) to which the business has access
COST	The business' ability to profitably capture the created demand

Successful oil companies will have to decide how to utilize these three levers to create unique capabilities for generating shareholder value. For example, time could be used to gain access to exploration opportunities before others, or the company could build platforms in half the time, reducing construction costs and accelerating cash flow from reserves.

Value could be leveraged by creating markets for otherwise uneconomic natural gas. Cost is important in all cases; the cost structure of the business must be low enough to capture value. Cost could also be used as a competitive edge, if a low cost position can be built and maintained.

Capabilities which leverage time, value, and cost must be built across the value chain (see Fig. 2–4) and extend beyond the company itself. Alliances may be required to fully exploit a competitive edge. (see Chapter 14 on the extended enterprise.)

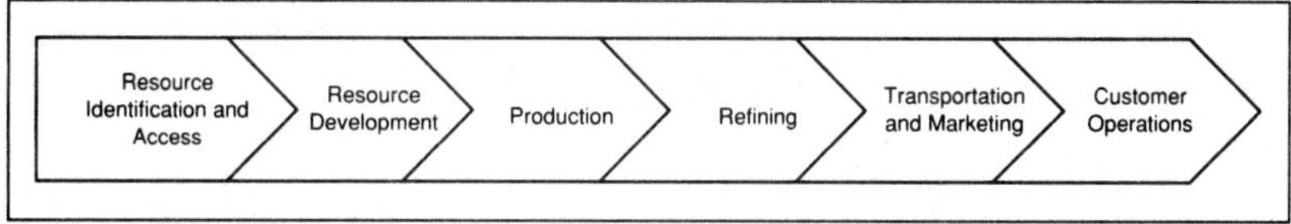

Fig. 2–4 Value Chain.

Market driving capabilities involve figuring out how to have better to compete with—not how to compete better with what you have. Market-driving capabilities:

- Are the business know-how that creates opportunities in advance of competitors
- Define competitive advantage—how you beat your competitors and create value for shareholders
- Go beyond "we're good at" and get to "how we will create value"

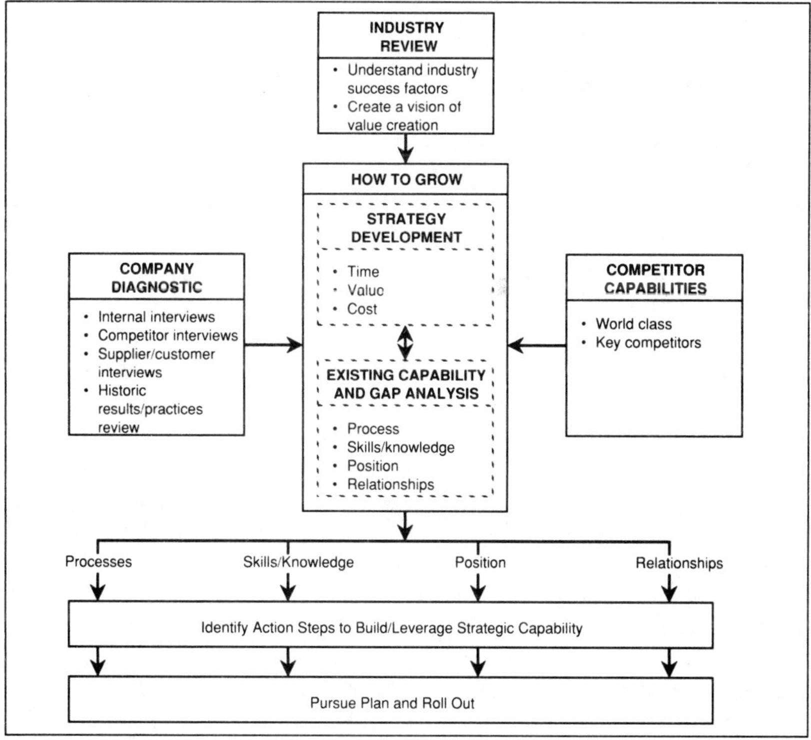

Fig. 2–5 Capabilities Framework.

Building a capabilities based strategy can be broken down into a six step process, focused not on "what will happen" but on "how to succeed" (see Fig. 2–5).

The key to a capabilities based strategy is that it is based on a set of skills, positional assets, processes, and relationships which, when combined, make up a strong and defensible advantaged position in the industry. These four components are defined as follows:

- ***Skills.*** Specialized technical, operational or managerial abilities focussed on critical aspects of the industry. For instance, these can include engineering, operating, research, and/or marketing abilities.
- ***Positional assets.*** Resources which give rise to a unique position, often difficult to replicate. These may include physical assets such as reserves position or production base, reputation or brand name in a region, and other related or unrelated businesses in a region that can be used as a critical mass in a region for starting up other businesses.
- ***Processes.*** Business processes for establishing and conducting business. Can include transaction and accounting processes, businesses development approaches, and labor management practices.
- ***Relationships.*** Business, government, and public relationships which allow the company to efficiently pursue its business. Can include joint venture partnerships, contractual or informal business relationships, working relationships with local and regional governments, and citizen advisory and participation organizations.

Examples of how companies have combined these to create winning strategies illustrate the power of a capabilities-based approach. While the previous examples have been drawn from the upstream, the same approach applies with equal validity to the downstream. Refineries, for example, can function as merchant refineries, selling their products into spot markets (the flexible model); they can supply a constant stream of products into a market with a low cost approach (the focused model); or they can produce speciality products for niche markets (such as lubricants). In each of these three cases, different capabilities will be needed. The merchant refiner will need to be a skilled operator and a quick decision maker to take advantage of market opportunities in the face of volatile prices. The focused refinery serving a large retail network may choose to develop a strong crude supply position linked to a refinery system structured to optimize around that specific feed and end product slate. The speciality refiner will need to be customer focused and abreast of the latest developments in his chosen end use market—perhaps as part of a strategic alliance

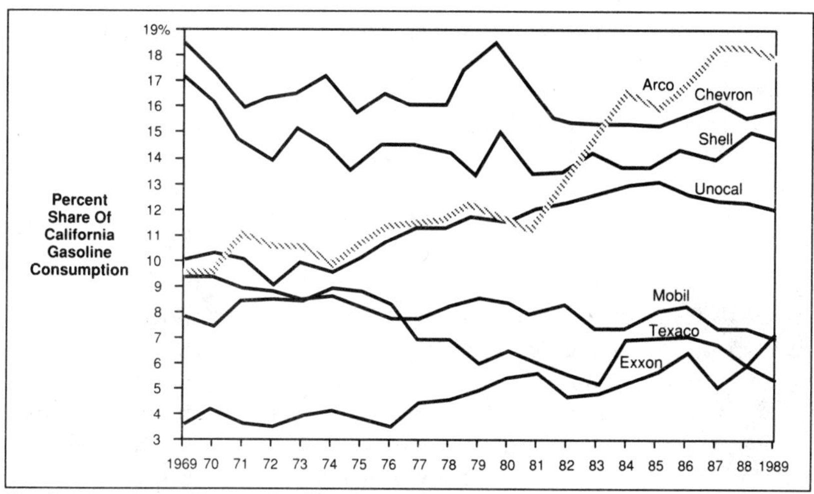

Fig. 2–6 Top Seven Shares of California Gasoline Consumption. Source: Lundberg Survey, Inc.; NPN data.

with retailers serving the end use market. Each of the strategies implies a different set of skills and organizing principles.

Examples of the development and application of market-driving capabilities downstream can be found in the recent strategies pursued by Enron, Arco, and Amoco. Enron has become a major gas company not only by amassing significant, low-cost reserves, but also by developing (or acquiring) the ability to monetize those reserves by entering the power generation business and by developing a comprehensive range of marketing and risk management services which complement and reinforce its position in gas markets. Many of these critical skills did not exist when Ken Lay created Enron. He and his management team set out to recruit and train the people needed to drive the emerging gas market. By all accounts they are succeeding admirably.

Arco chose another course—that of geographic concentration—and undertook a radical restructuring of its asset portfolio to focus its downstream operations on the U.S. West Coast. Arco built a low-cost value chain which brings its large Alaskan oil supplies to the major transportation fuel markets of California, Arizona, Nevada, and the Pacific Northwest (see Fig. 2–6). As a result, Arco's market share climbed at the expense of higher cost competitors.

Amoco chose yet another course. Building brand value in both gasoline and diesel fuel, the Chicago-based company capitalized on its marketing strengths while other companies failed to develop comparable marketing skills. Amoco consistently receives more for its fuel than most—if not all—of its competitors. The capabilities Amoco developed where those typical of a successfully retailer. The company used advertising effectively and consistently, while keeping costs under control. The results speak for themselves: Amoco has enjoyed a strong record of downstream profitability while others earned nothing but red ink.

CONCLUSION

n order to succeed, companies must move beyond the notion of strengths and weaknesses and begin focusing on building and deploying capabilities to success. These capabilities should be measurable. Strategies in the twenty-first century will focus less on asset acquisition and more on building distinctive capabilities both through in-house efforts and extended enterprises. These capabilities will pertain as much to how the company functions (flexibility, decision-making speed, and so forth) as they do to technical excellence.

CHAPTER 3

KEYS TO UPSTREAM SUCCESS

Successful management of the upstream oil company requires much more effective linking of key decision processes than is traditional in the industry. As shown in Figure 3–1, value can be created in the upstream in four basic ways:

- Add commercial reserves by drilling at low cost relative to the competition.
- Produce reserves at high margins relative to the competition.
- Add commercial reserves by acquiring assets at low cost relative to the competition.
- Monetize reserves by selling assets at a higher price than you value them.

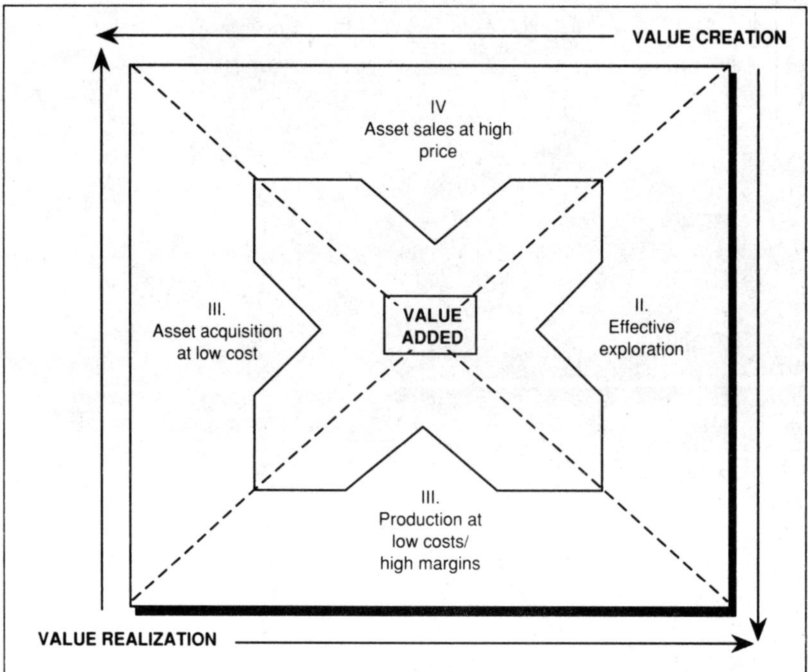

Fig. 3–1 Components of Value Creation.

These four ways to add value can and should be managed as a portfolio of opportunities. We call the process of doing so "managing the opportunity funnel." But each also has its own set of key success factors. We will describe first how the four are linked and then suggest the fundamental differences among the capabilities required to succeed in them.

THE OPPORTUNITY FUNNEL: MANAGING THE ROLE OF SELECTIVITY AND LEARNING

ithin every oil company, the game of adding value begins as a competition for capital. For example, exploration project teams compete with enhanced recovery groups, and geographies vie with geographies for scarce capital and human resources.

This process has worked well within the industry over the years, because it has resulted in successful prioritization, or "highgrading" of projects. There are at least two pieces of evidence of this success. First, on average, the largest and most attractive reservoirs of oil and gas have been discovered and developed, before less attractive pools. In North America, the world's most mature area, 35 fields greater than 500 BCF were found in the 1960–64 period, but only three were found in a similar time frame in the late 1980s, with a relentless decline in between, other than the peculiar early 1980s with their price distortions (see Fig. 3–2). When we look at three separate producing provinces, in each at least 80% of the hydrocarbons have been discovered from 20% of the wells (see Fig. 3–3). Oil companies seem to set their priorities well.

A second piece of evidence is that reserve addition costs tend to rise as the rate of reinvestment in drilling and acquisitions rises. That is, companies that plow back into reserve additions a greater proportion of their E&P

YEAR ROUND	NUMBER OF FIELDS	TOTAL LIFETIME GAS YIELD (TCF)
60 – 64	35	39.4
65 – 69	22	33.5
70 – 74	11	9.0
75 – 79	5	3.4
80 – 84	9	7.2
85 – 89	3	3.7

Fig. 3–2 North American Gas Field Discoveries 1960-89. Source: Dwight's, BA&H analysis.

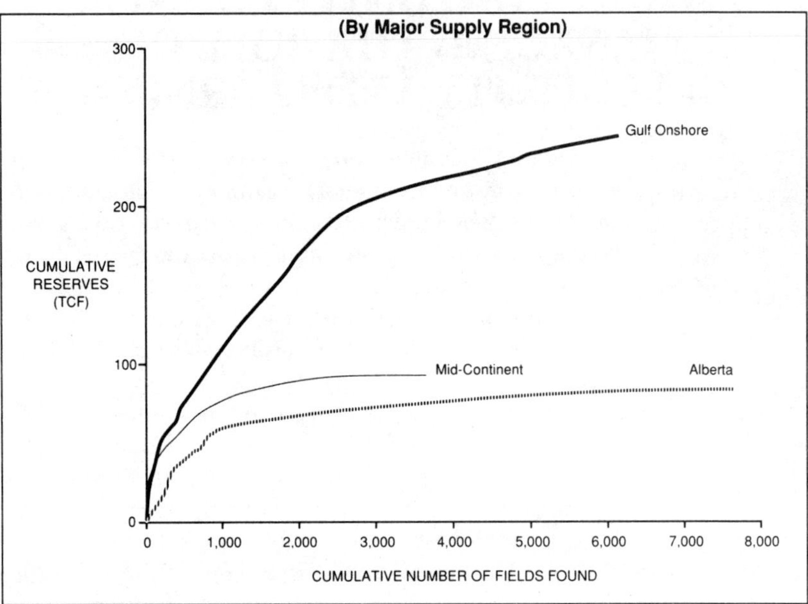

Fig. 3–3 Total North American Gas Reserves Found Verses Number of Fields Found. Source: Dwight's, BA&H analysis.

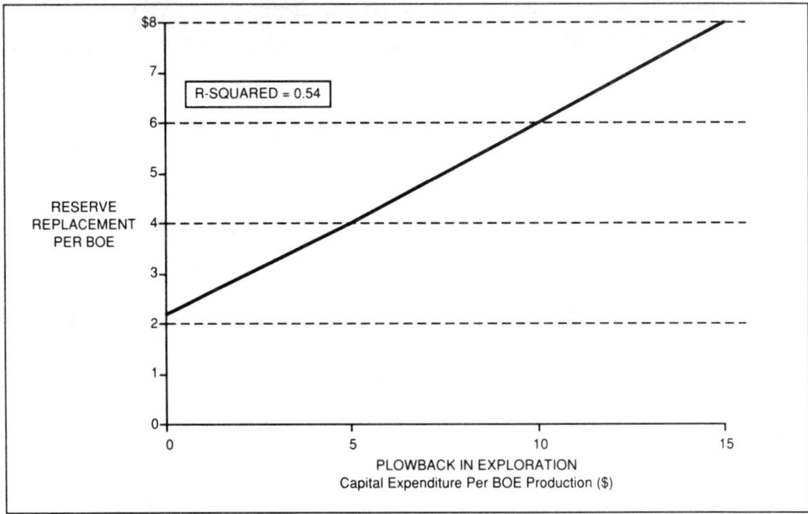

Fig. 3–4 The Relationship Between Plow Back Rate and Reserve Addition Costs. Source: J.S. Herold, BA&H analysis of ten major companies.

cash flow tend to incur higher costs. (Note that in Figure 3–4, annual hydrocarbon production in barrels of oil equivalent is used as a proxy for E&P cash flow to keep the data consistent.)

As the figure shows, rate of reinvestment explains 54% of the variance among the reserve replacement costs of oil majors.

Why is this, and why is it important? We believe that inherent in every company is an internal limit to profitable growth, a limit which is represented by the scale of "resources and reach" of the company. This makes sense because one would expect a company that consistently invests $4 in exploration capital per barrel of production to be highgrading its opportunities more than a company which consistently invests $8.

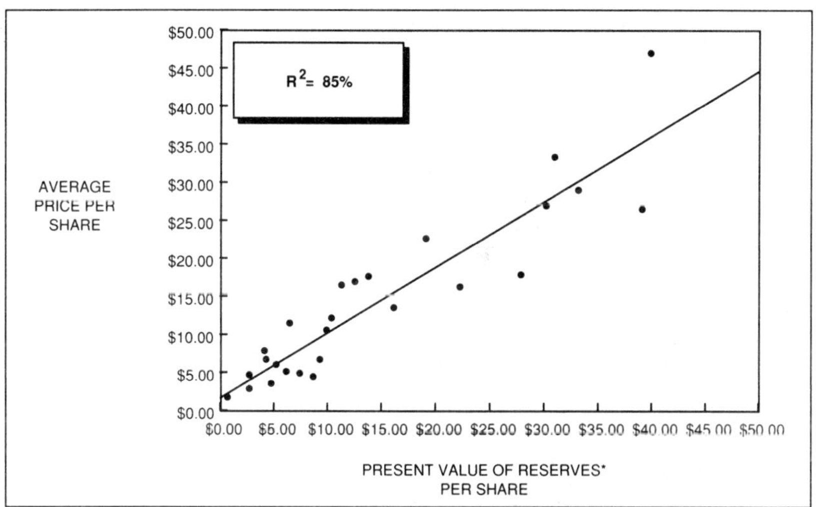

Fig. 3–5 Future Cash Flows from Existing Reserves of E&P Value. Source: Derived from J.S. Herold's "Appraised Net Worth" calculation—J.S. Herold; BA&H Analysis.

This point is important because companies that attempt to grow their reserve bases very quickly are likely to perform worse in financial markets than those that invest steadily and more conservatively. Capital markets value upstream operations on the basis of the cost incurred to add proven reserves in the ground

Lower costs of addition mean higher market values, and strong highgrading leads to lower costs of reserve additions. There is therefore a relatively efficient market for investment opportunities, which again is evidence that highgrading works, at least for exploration opportunities (see Fig. 3–5).

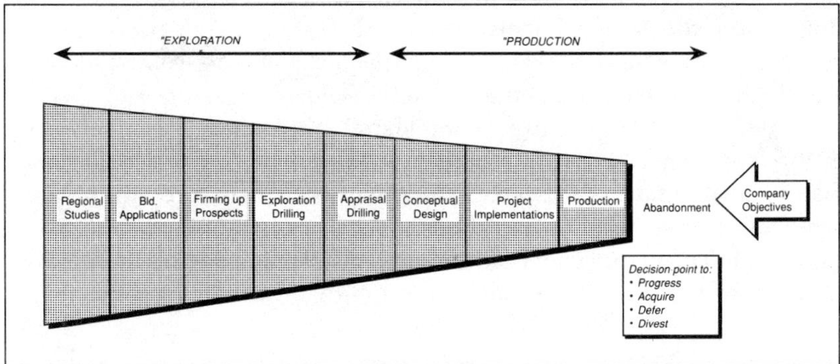

Fig. 3–6a Opportunity Management Funnel: Activity Steps.

But rigorous and consistent comparison of investments is getting harder to do, as companies expand their search for value away from their traditional hunting grounds. Beginning in the 1970s, major private oil companies began to look for and to produce reserves far afield from where the industry began. Nationalizations in the Middle East and elsewhere and the relentless maturing of North America as an oil and gas province, drove this

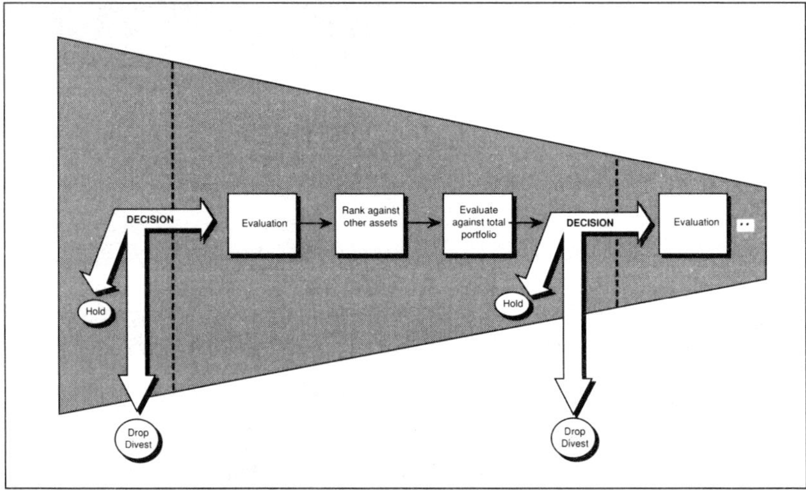

Fig. 3–6b Opportunity Management Funnel: Decision Steps.

expanded search. For example, the ratio of E&P expenditures in the United States vs. elsewhere inverted itself between 1985 and 1993.

This rapid redeployment means that people and technology are becoming ever more widespread, and that opportunities are getting increasingly exotic and hard to compare. Executives are constantly facing challenges such as whether to invest in oilfield turnarounds in Russia, or a new field in Viet Nam. How can a company adequately compare the risks and returns of such different opportunities? Is there a framework that might help with the selection of investments?

We began to develop such a framework in the late 1980s. We call it the Opportunity Management Funnel. It is a framework which allows management to evaluate and select opportunities among the four value addition options discussed above. Figure 3–6a illustrates the funnel. Each segment of the funnel represents a company's unrisked hydrocarbon potential in that piece of the chain.

The ideal shape of the funnel for a specific company is determined by the company's objectives (such as production per year or economic value added per year) and an assessment of the company's (or the industry's) experience. Once this ideal funnel shape is defined, the company can compare the ideal with the actual portfolio of opportunities the company owns. There usually are major gaps between the ideal and the actual. For example, a company may be thin on risked reserves in licensed predrill prospects, and abundantly endowed with operating field assets. If so, the comparison of funnels would direct management to shift resources to improve the company's license position, or to change its objectives.

A second analysis could compare the funnels for different regions within the company to see where the shortages and surpluses of assets in the portfolio are located. This analysis would result in a shift of resources among regions to address the differences between the company's funnel and the ideal shape of the portfolio.

We have found that the information and processes required to manage an upstream business using funnel principles are rarely in place. Most companies are surprisingly unsystematic in how they look at worldwide opportunities. However, once the processes are in place to manage this way, the portfolio should, in time, come into line with company objectives and should help lower asset-building costs.

It is important to note that assets can *enter* or *exit* the funnel at any segment in it by acquisition or divestment, which are, of course, two of the four ways to add value mentioned in Figure 3–1. Drilling and production are represented by movement through the funnel, once an investment is made (Fig. 3–6b).

KEY SUCCESS FACTORS

aving discussed a framework for thinking about portfolio management, let us turn our attention to nine factors which, if built into your company, will help you to be more successful. These are by no means the only factors of success, but they are the themes that arise again and again from our work with clients.

CROSS-FUNCTIONAL KEY SUCCESS FACTORS

From the discussion above, we can identify two key success factors which apply across-the-board for the upstream:

1. Understand your target and actual portfolios of assets.

Know your objectives and how these objectives should affect the shape of opportunities you have in your portfolio of assets. Have the information on your business in a form that makes the funnel understandable and manageable. We have found that most upstream companies do not have their asset information in shape to do this.

2. Redirect resources to close gaps in the portfolio between target and actual.

This is the hard part. Once you know where your strengths and deficiencies are in the funnel, you have to redirect people, capital, and technology to the areas of deficiency. For many companies, this has meant major redeployment.

3. Be more 'selective' than your competitors.

Upstream oil companies, by and large, are on a level competitive playing field. They get their people from the same schools, and all strive to get the best people. Advanced technology is accessible to all companies, although one company might sustain a technological edge for a short period in a particular application. Similarly, capital is available to well-established companies under similar conditions. Over time, therefore, the primary distinguishing feature among upstream companies is how selective they are in upstream investments. The more rigorous the highgrading, the better. And no company should base a long-run strategy on catching up to its competitors by high investment rates relative to its competition.

KEY SUCCESS FACTORS IN EXPLORATION

We will focus in this discussion on the key success factors for wildcat exploration, which we will define as the search for "elephant fields."

True frontier exploration investment is closer to pure research, or to venture capital, than it is to traditional business where the risks and returns are reasonably well known. The products of pure research are notoriously hard to predict, and venture capital is more littered with failed investments than with great successes. But these research/speculative investment activities go on in every industry, and in fact, the leaps in competitive advantage from successes in speculative investments are becoming more important as competition intensifies in industries such as pharmaceuticals, software, and technology, in addition to oil and gas.

The tremendous efforts many companies go through to compare true frontier opportunities with step-out or developmental plays often is not appropriate. Hurdle rates and discounted cash flow comparisons cannot shed much light on this process. Frontier exploration requires a venture capital approach to funding. Capital needs to be set aside with the knowledge that it will not pay out at any predictable time, but over the course of a decade one would expect a major discovery yielding returns far in excess of the investments made.

We therefore suggest another two additional key success factors which apply to world wildcatting:

4. Do not overrely on quantifiable data.

Instead, pick your best "discoverers" and apply them over the long run to use instinct and experience to come up with new elephant prospects. Measure them by their ability to compete with other opportunities, but take them out of the annual budget, short-term competition that most other opportunities require.

5. Establish the capability within your company to know when a method of exploring in a basin has played itself out, experimenting with new approaches, and leaving when it makes sense. This means understanding the 80–20 rule illustrated in Figure 3–3.

DRILLING AND PRODUCTION KEY SUCCESS FACTORS

As an asset moves down the opportunity funnel, risk management gives way to cost management as the preeminent concern. Assets turn from ideas or prospects to drilling pipe, field infrastructure, and proven reserves. The key success factors of production play to the core capabilities of an often different group of companies than those that lead the way in exploration success.

Perhaps the most important key success factor for production is the following:

6. Get the most out of partnerships, outsourcing, and the like.

Many oil companies have gone through the exercise of defining their core capabilities to decide what they are particularly good at. Benchmarking of various kinds has helped to illustrate that most companies are outstanding at a relatively few aspects of drilling and production, and that these strengths differ greatly among companies. This fact, combined with the need to lower costs and improve margins, is driving the industry toward a proliferation of new combinations in the form of familiar joint ventures and partnerships, to alliances and creative contractual arrangements. The aim of these combinations is to match one company's core strengths with another's weaknesses, and vice versa.

This is the micro application of the concept of comparative advantage: we are all better off if we do what we do best and rely on others to provide what we are not very good at providing. Most major oil companies—BP, Arco, Texaco, etc.—have left or greatly scaled back their onshore U.S. operations because their cost structures are simply too high to compete in small fields and because their comparative advantage lies in bigger, riskier projects that play to their technical and capital formation strengths. Similarly, Occidental sold its Argentine production interests to domestic groups because the properties could not shoulder the costs of ex-patriates, and because Occy is an explorer by inclination and wanted trade cash for value elsewhere.

Most joint ventures, mergers, and alliances are evidence of two companies relying on one another for strengths they themselves do not have. In the North Sea, partnering is taking on new dimensions as large companies, continuing to streamline all aspects of their businesses, are outsourcing more and more platform design, development, and production activities to low-cost providers. Accounting is outsourced as well.

Research and development (R&D) provides further examples of the role of core capabilities and combinations. Most oil companies now recognize suppliers as major sources of innovation in many areas. Innovations geared towards lowering costs in development in the upstream sector are being made by joint development teams comprising oil companies, equipment suppliers and design companies, each bringing their specific capabilities to bear. Subsea manifolds and unmanned satellites are examples of

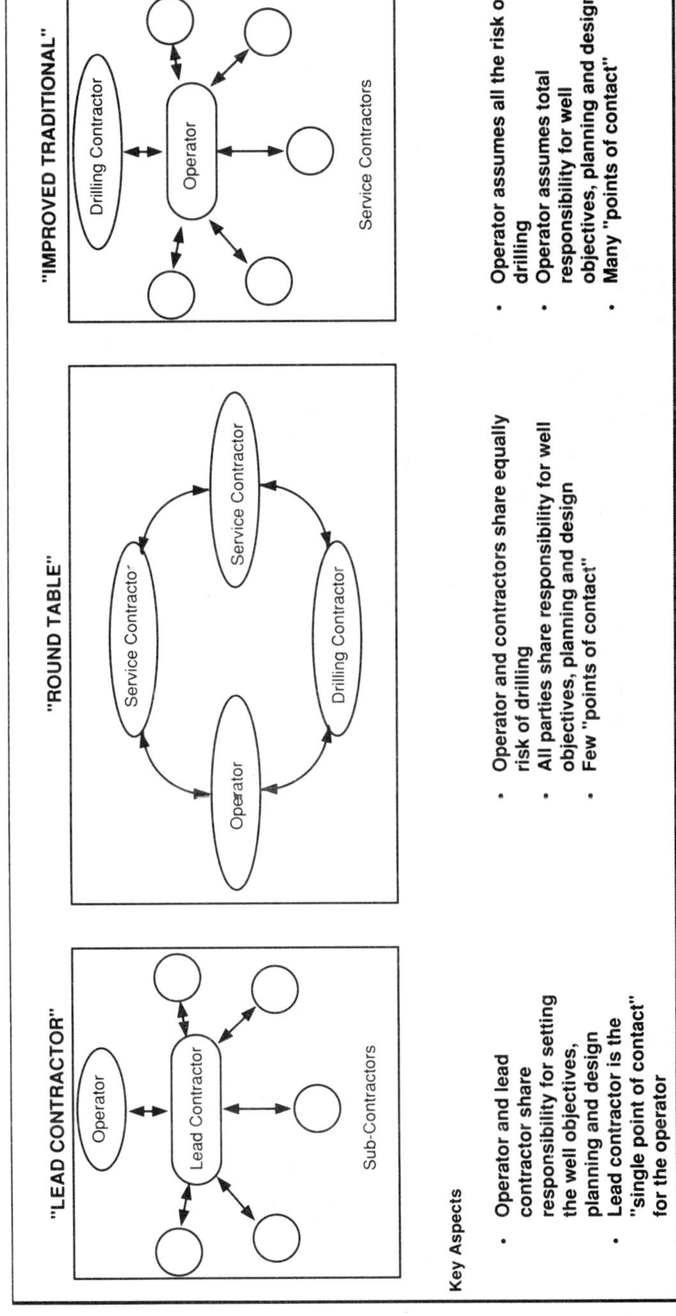

Fig. 3–7 Preferred Contractor/Operator Team Relationship

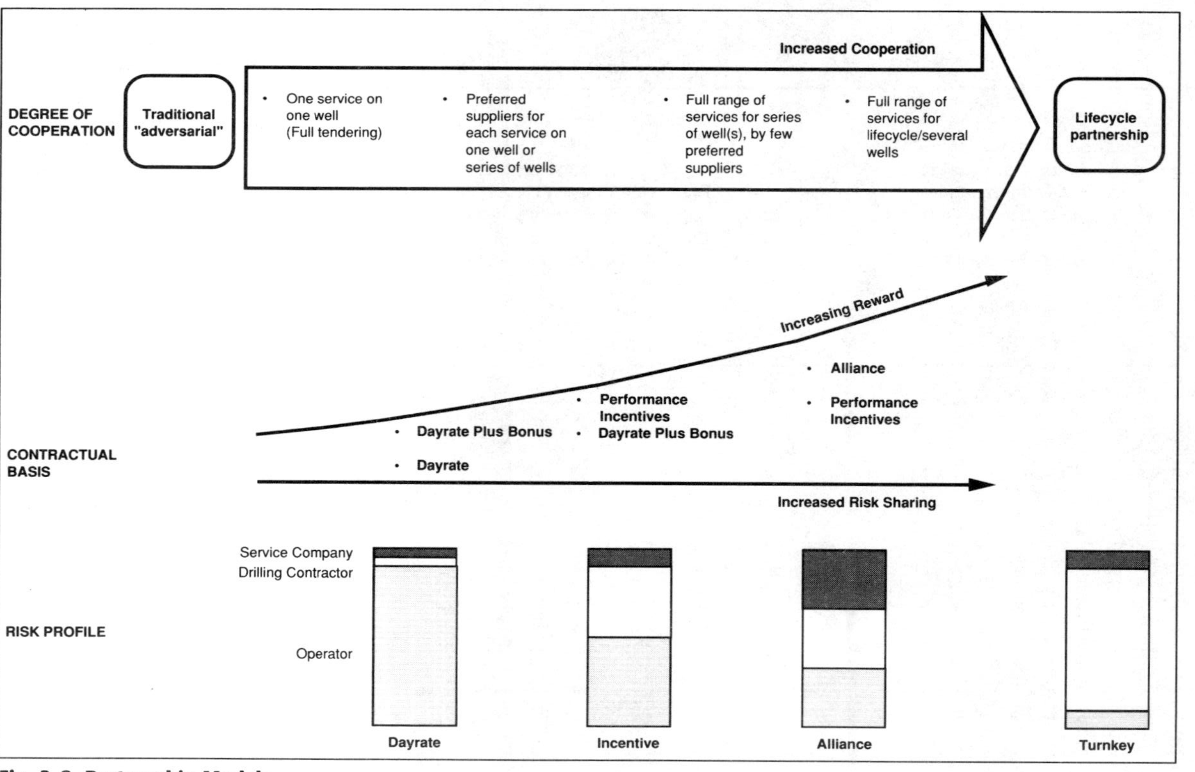

Fig. 3–8 Partnership Models.

strong partnering. The significant advances made in the applications of computing technology are another.

The challenge has therefore become one of how to apply new technology developments effectively. For new field development, companies are moving at different speeds towards implementing the use of international standards for design criteria and the use of functional rather than company specifications for equipment purchases. Much work has been done on these topics, e.g., a joint industry team (CRINE—Cost Reduction in a New Era) is researching new ways to reduce costs in the North Sea. The initiative has been well received, but the pace at which individual companies are willing to implement recommendations varies from company to company, depending on their history of using in-house developed standards and specifications, inclination of existing functional departments to protect their turf, and other factors. We would argue that the most successful companies will be those able to apply international standards and functional specifications to meet the right level of "fit for purpose" quality in their operations around the globe.

Outsourcing of detailed design, procurement, and construction management is now commonplace. However, there are still large differences in the size and type of project teams involved and the form of contractual arrangement for the work. Project teams can range from operators with 100–200-man independent project teams to small fully integrated (operator, design company and contractor) teams where it is impossible to distinguish who actually works for the operator, the designer, or the contractor—each project team place being filled by the "best person for the job." We would argue there is a significant cost to be paid by the operator who awards and administers many individual contracts and chooses to police his contractors. Some of the different approaches are shown in Figures 3–7 and 3–8.

A new approach is emerging based on a more co-operative "win - win" contracting approach in which the goals and objectives of all the parties are shared and aligned. In future, contracts will be based on longer term commitments, have fewer or no adversarial clauses, provide incentives for excellent performance and involve less manpower/interfaces in their preparation. Examples abound of savings in the order of 10-25% on original capital expenditure and schedule if these type of incentive based alliance and partnering relationships are implemented. Whatever the exact contractual arrangement, fundamental to success will be to develop a culture of mutual trust, objective critique and open communication between all parties. This will not be an easy task given the historical approaches to project development around the globe where "policing" of the contractor was very much the norm.

In operations there has also been a clear shift to outsourcing of low value added activities. The more progressive operators are taking more of a lifecycle approach to maintenance activities. The actual maintenance repair work was historically performed by the operator himself while today much of the low value added day to day repair work is outsourced to a maintenance contractor with the operator focussing on planning preventative and predictive maintenance, organizing contracts and supervising the work. In future we expect there to be a growth in the number of suppliers who will be obliged not only to deliver their equipment but also to install and maintain it (including stocking of spares) over the lifecycle of the particular development. This is already the case in many areas of the world for drilling equipment (wellheads and BOPs) and pumps.

KEY SUCCESS FACTORS IN ACQUISITIONS AND DIVESTMENTS:

Two of the four value creating activities are not usually mentioned in the same breath as the ancient twins of exploration and production. But in fact, acquisitions and divestments can be of equal importance in determining value to shareholders. As we have seen, perhaps the most important decision that faces an upstream company is when to sell assets to a company that values them more highly. Similarly, acquisitions offer the possibility of great wealth creation or destruction.

These two value creation mechanisms must be core skills of most oil companies. The reason is that they require careful evaluation of a company's assets and its competition and/or potential buyers and sellers. These activities tend, therefore, to open up the culture of a company and to build its commercial capabilities. These features, while "softer" than drilling or production skills, are fundamental to adding value to a companies owners.

Key success factors in this area follow:

7. Do not covet your own assets.

Look at every asset as an opportunity to provide value through any one of the four means in Figure 3–1. At each stage in the opportunity funnel, decide whether to hold the asset or divest it. Deflect arguments from management and staff that the asset is "part of our history," "part of who we are." No oil company can afford that kind of reasoning. If it means so much to us, why is it that someone else values it more highly?

8. Build and manage an asset acquisition/divestment information base.

In our experience, most oil and gas companies know a great deal about themselves and their competition, but this knowledge is contained within individuals. Institutional memory is lacking. That is, very little insightful and timely information is easily and quickly accessible for the purpose of valuation. This database and analytical capability is fundamental to management of the funnel.

9. Attract and develop negotiators and financial engineers with operational knowledge.

Every oil company needs these people and plenty of them. Unfortunately, these are extremely rare people. They usually cannot be gotten straight out of school, and it takes many years of concentrated development to build them in-house. Most oil companies do not value or even recognize the type, confusing them with investment bankers or planning people. The best acquisition/divestment people have deep operating experience and a knack for numbers and dealmaking.

CONCLUSION

he upstream oil business has traditionally been defined by differences in its various disciplines, geographies, and risk profiles. So we have long heard of the territories of the geologists vs. those of engineers vs. those of financial people, and so forth. These attitudes toward organization have led to many internal barriers that many companies are still trying mightily to break down. And most companies have been characterized by substantial pride in themselves, leading to overconfidence and a building of barriers to the outside world.

We have suggested here that a much stronger orientation toward managing the entire portfolio of assets is required by understanding your opportunity funnel and rebalancing it as required, regardless of whether an asset is oriented toward exploration or production. At the same time, the key success factors direct companies toward more open and innovative management of partner relationships, essentially requiring opening the borders of the typical upstream company to immigration of new ideas and toward greater outsourcing.

Such changes require fundamental changes in the organization, culture, and individual talent and behavior of employees. These topics are addressed in subsequent chapters .

CHAPTER 4

HIGH PERFORMANCE REFINING

Refining is the critical technological link between the upstream and the downstream. Crude oil is refined to produce the wide range of petroleum products that consumers want (small quantities of crude are directly burned in Japanese electric utilities). Yet refining organizations in most integrated companies have only recently come to be viewed as stand-alone businesses.

REFINING STRATEGY AND OPERATIONS

he starting point for any refining strategy must be the existing physical configuration of the company's refinery or refineries. Refineries range from small and simple topping units that boil crude oil into a very simple slate of basic products to extremely large plants that produce a wide variety of end products.

However, as illustrated in Figure 4–1, there are three basic types of refiners—complex, cracking, and hydroskimming—found in most developed country markets today. They differ dramatically in their ability to produce lighter products (e.g., gasoline, diesel) at the expense of heavier products (e.g., residual fuel oil).

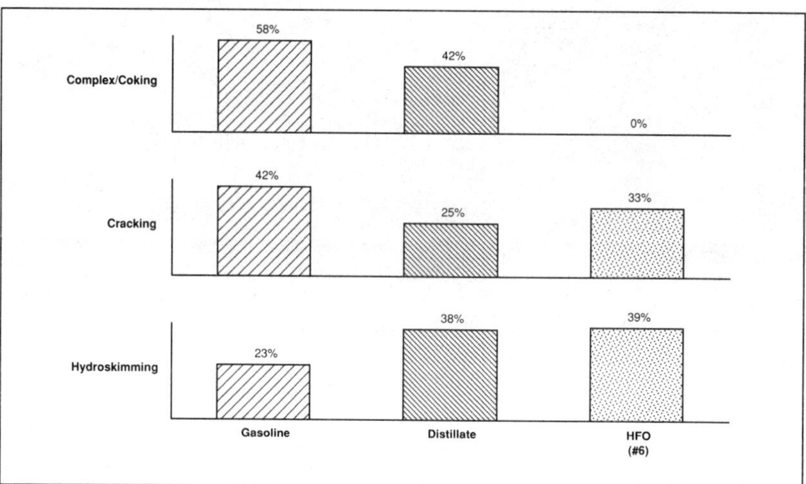

Fig. 4–1 Yield Structure by Capacity Type.

For each refinery, the desired set of products is one starting point, while the crude selection process is the other. In reality, these two are inextricably linked and should be considered jointly. The goal for any refiner should be to maximize uplift across the refinery. Determination of the appropriate refining strategy is an iterative process, balancing the changing product yields with the changing crude slate to optimize the dollars per barrel uplift.

All refiners must decide whether they can make money being a low-complexity/low-cost refinery or a high-capital/complex refinery with a large

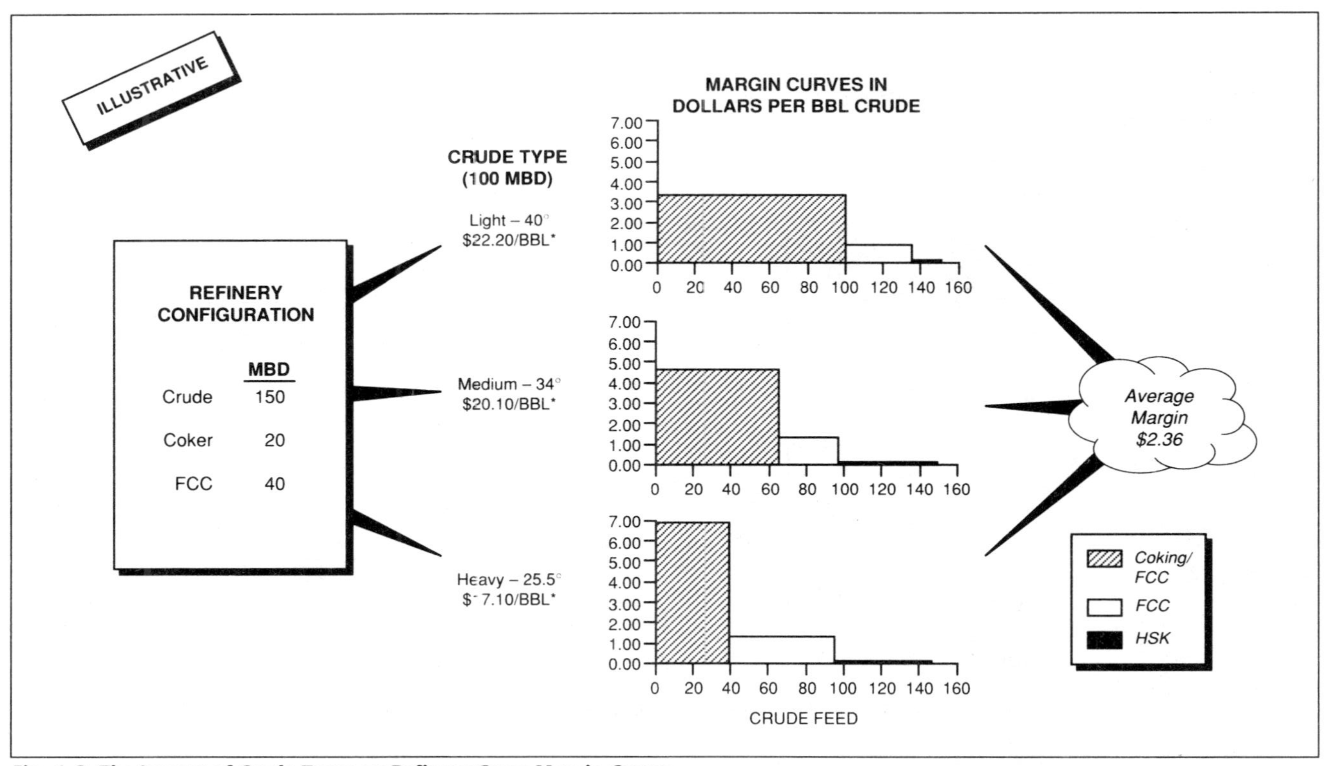

Fig. 4–2 The Impact of Crude Types on Refinery Gross Margin Curves.

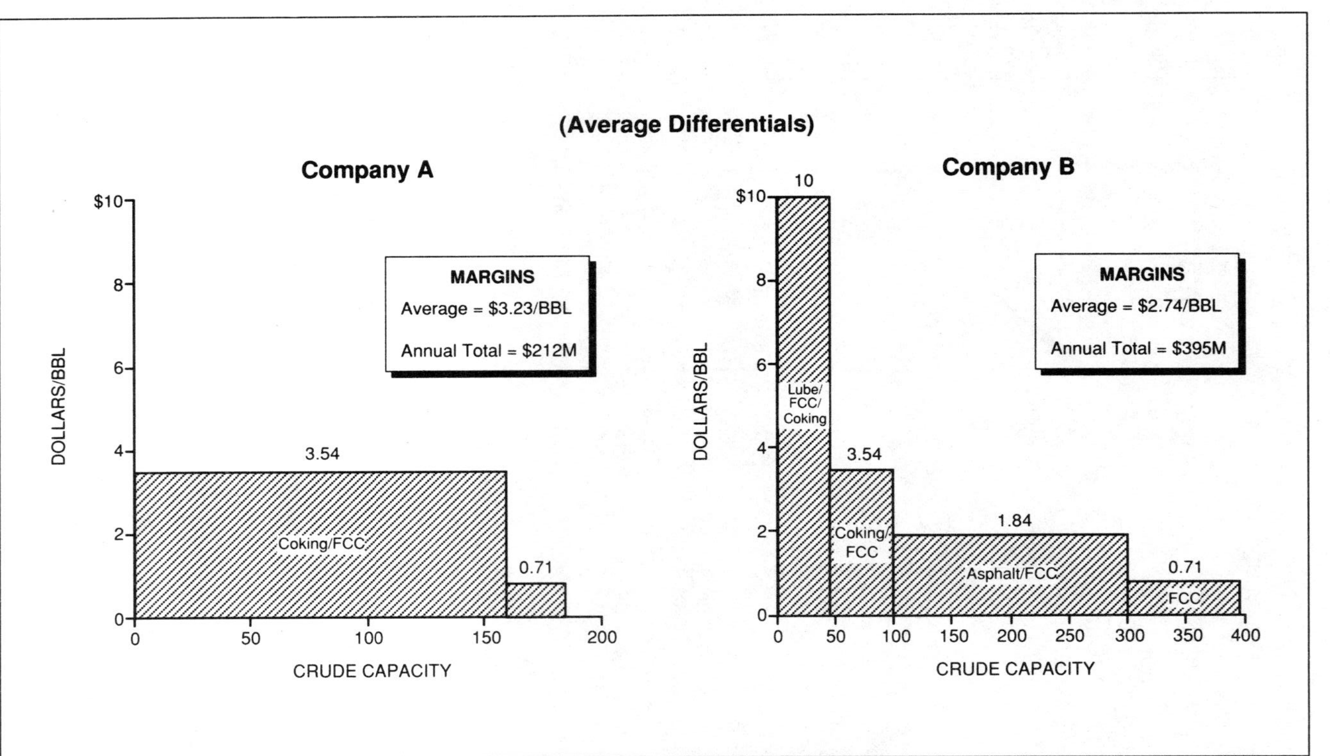

Fig. 4–3 Refinery Gross Margin Curves. Sources: *Oil & Gas Journal*, Pratts oil grain prices, BA&H analysis.

gross margin sufficient to support the additional debt load. Some refineries do not have a choice: they maybe simple refineries without access to capital and therefore have to be as innovative as possible with a fairly constraining set of assets. Other complex refineries already have committed the capital and must continually look at a variety of crude and other feedstocks and new ways to best use a more flexible set of assets.

The following examples are designed to illustrate the complexity of refining strategy choices.

For a given configuration, different crudes will produce a predictable yield of products, as shown in Figure 4–2. As the product yield pattern changes, the gross margin curve will also change.[1]

Strategies must therefore contend not only with the structural differences between refineries, but also with the effect those strategies will have on the markets being served. As shown in Figure 4–3, different refineries have inherently different margin curves. The chart shows how Company A enjoys a margin of $3.23/bbl while Company B earns only $2.74/bbl.

However, operational decisions within given configurations can also have significant impacts. As shown in Figure 4–4, companies with similar processing units (in this case, a reformer) do not always run those units at optimal levels. The chart plots the "lost opportunity" of running the units at full capacity.

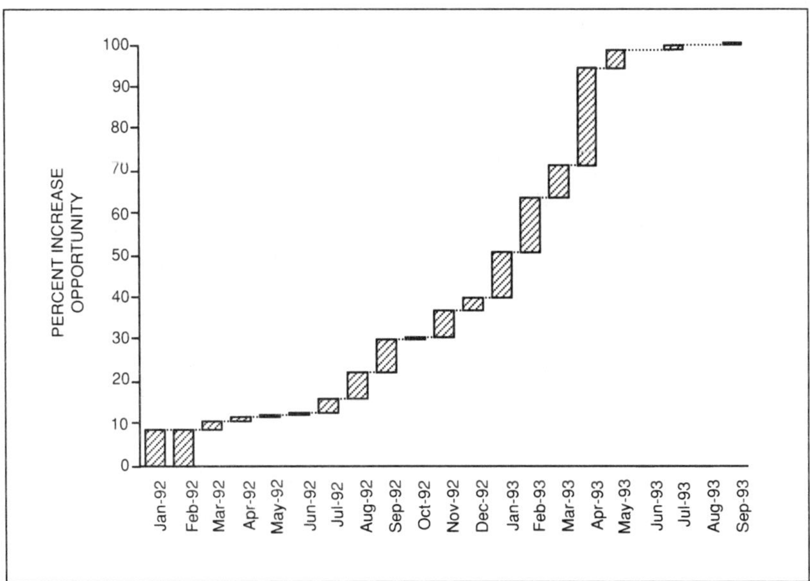

Fig. 4–4 Refinery a Reformer Utilization Opportunity.

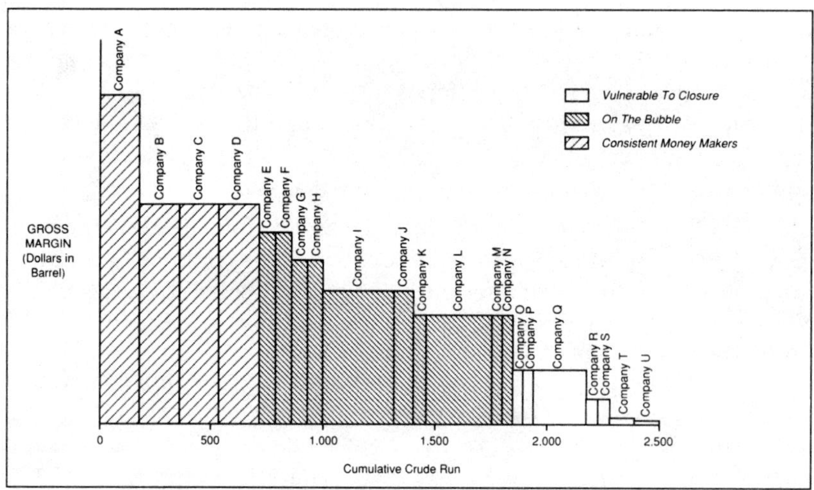

Fig. 4–5 Regional Refinery Margin Curve.

For a given refining region, it is in fact possible to construct an aggregate refining margin curve, as shown in Figure 4–5. Of course, this chart is an oversimplification, since it only examines the average margin for all products. A more sophisticated approach would measure the marginal supply curve for each product.

Only with a thorough understanding of a refinery's position in such a regional market can a manager make strategic and tactical decisions.

If you are fundamentally disadvantaged—for example, if you work for Company R in Figure 4–5—you are unlikely to turn a consistent profit unless demand growth and margins are robust over a prolonged period of time. You should therefore consider if major improvements in your assets or in operational performance are possible or whether it would be better to leave the market. On the other hand, if you work for Company A shown in the same figure, you are in an enviable position and should be able to profit under virtually any market condition.

A key activity for senior refinery management is continuously looking out into the future and assessing the impact of known and possible future events. For example:

- What impact will new coker capacity have?
- What will the introduction of reformulated gasoline do?
- Will a new pipeline suddenly allow additional refineries to market products in our region?

BUILDING A HIGH PERFORMANCE REFINING ORGANIZATION

The best strategy in the world often falls victim to poor execution. Development of a high performance refining organization is therefore a critical companion to your strategy and requires a three-dimensional perspective. Specifically you'll want to know:

- How refining should relate to the other segments of the company—for example, upstream, marketing, and chemicals.
- How to run an individual refinery.
- How to run a portfolio of refineries.

While there are a few refining companies in the world which are totally independent of other upstream and downstream activities, the vast majority of the world's refineries are owned and operated as part of a vertically integrated enterprise. The interaction of refining with crude production, on one hand, and with products marketing, on the other, is no trivial pursuit (see Fig. 4–6).

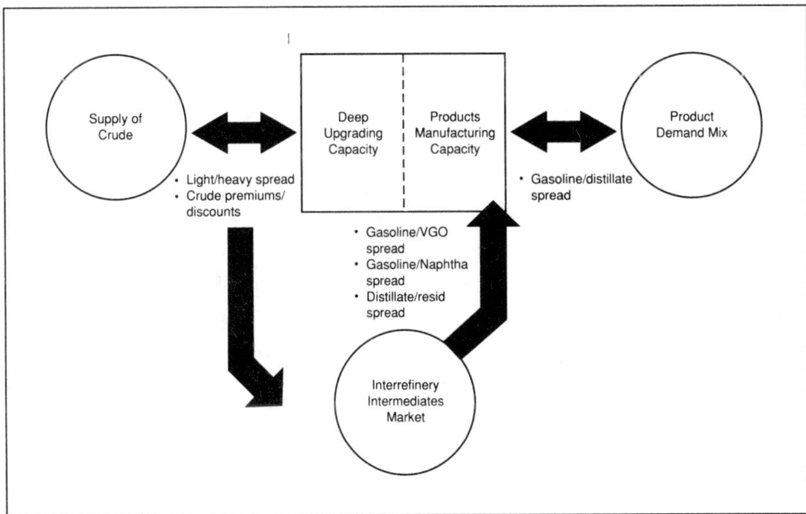

Fig. 4–6 Refineries in the Value Chain.

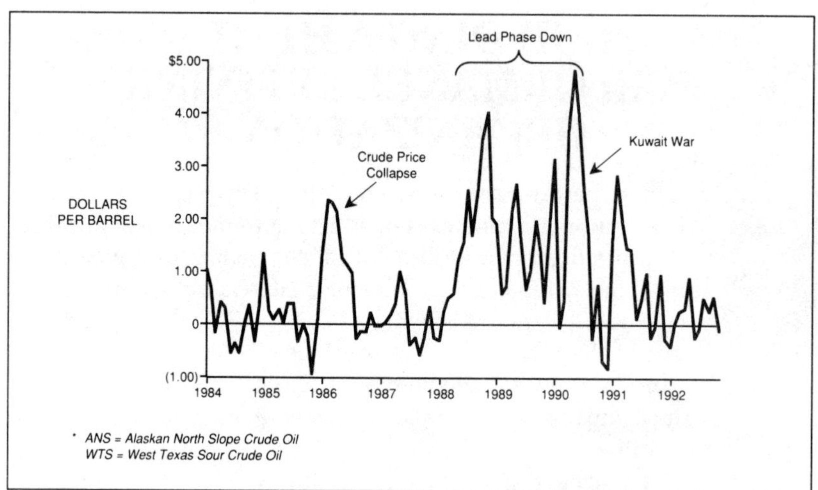

Fig. 4–7 ANS/WTS* Gulf Cracking Margin. Sources: PIW; Platts; BA&H analysis.

The increased volatility of prices and therefore of the spread between crude and product prices can swing refinery profit margins from red to black in a nanosecond. (see Fig. 4–7).

There is no guarantee that margins will be positive. As illustrated in Figure 4–8, hydroskimming margins in the U.S. midwest have often been negative.

The supply function, discussed more completely in Chapter 5, therefore becomes critical to the optimization of refinery assets.

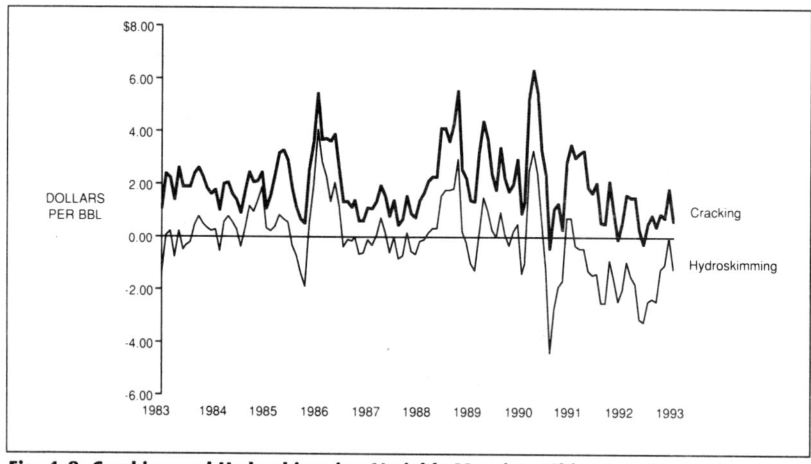

Fig. 4–8 Cracking and Hydroskimming Variable Margins—Chicago Area. Sources: Platts Oil Gram Prices, BA&H analysis.

However, in any company, the intersections of different business units or functional departments give rise to tensions and inefficiencies. The intersection of refining and supply is no different.

A critical task for senior management is to ensure that the whole is not being suboptimized at the expense of optimizing the individual business units: refining, marketing, chemicals, etc. Performance measures and reward structures must be in line with maximizing the profitability of the overall corporation and not providing incentives to focus only on the local operation.

Even where divisions are closely integrated on the same site , such as a refinery, petrochemicals and possibly natural gas liquids operation, individual objectives can supersede overall company goals. Realistic and frequently reviewed transfer prices are the key to maintaining the right operation. Where integrated operations occur, it is desirable to have one linear programming model (LP) that optimizes the total site, even if managers have to argue later about how much of the profit should be allocated to their part of the operation.

HOW TO RUN AN INDIVIDUAL REFINERY

There are eight critical success factors that generally apply to the management of an individual refinery (some also apply across refineries). Apply these best practices consistently, and you will have a high performance refining organization.

1. ***Listen to the market.*** Price volatility can swamp refinery economics. If you are losing money, you can't make it up on volume . In both crude buying and product sales, you must stay carefully tuned to the market signals.

2. ***Optimization continuously:***

- Refineries must have a good LP representative of the operation and run soon after crude or product price changes.
- Once the crude throughput and mix has been determined, individual unit models should continuously optimize each unit.

3. ***Organize around asset or area teams.*** Functional silos in refineries are legendary fiefdoms. Maintenance, operations, and engineering are often sovereign kingdoms with inefficient and ineffective interfaces. Organizing the refinery around cross-functional work teams will break down these barriers, thereby reducing costs and improving reliability.

4. ***Trust the operators.*** They know the equipment best, and they want to do a good job. The problem is they often lack the tools to make the right decisions. If you give them the right information and training, they can work wonders.

5. ***Stay flexible.*** There will always be peaks and valleys of work. Your organization should have the flexibility to share resources in order to conquer the peaks and to take advantages of the valleys. This is particularly true in the maintenance and engineering functions.

6. ***All maintenance is not equal.*** Refineries require a lot of maintenance, but determining how much by whom to what units and when can make a big difference to the bottom line.

7. ***Energy is not free.*** Surprisingly, many refineries do a poor job of tracking energy costs—particularly steam—by unit.

8. ***Build project discipline.*** Refinery turnarounds are major events and cost big bucks. The downtime is often the most expensive component of cost. Refinery managers should insist on careful advance planning that minimizes downtime and leverages the organization's skills in procurement, logistics, and doing the work. Operators should play important roles in turnarounds along with contractors.

Let's look at each of these in more detail.

Listen to the market

The critical linkage here is between the supply and trading organizations (which may be under the same roof or in another city and/or in another division of the company). Crude and product prices are constantly changing, with profound implications for refinery profitability. The cycle time for making crude purchase decisions is critical. Traders will only offer their best deals to companies that will respond quickly. The best companies can make crude substitution decisions in less than an hour. The worst companies take days or weeks and often refuse to deviate from quarterly plans. The same pattern holds true for modifying the product slate on make vs. buy decisions.

Far too often, refineries have been run on the basis of quarterly plans which make assumptions about crude and product price differentials (because of the time span involved those forecasts are guaranteed to be wrong). Information systems can help expedite this decision making by providing traders with accurate and timely information and analysis.

Optimization

A good linear program is at the heart of any well optimized refinery operation. It must be kept up-to-date and properly reflect the unit by unit operation. As new data on different types of operations and new crudes are gathered, new LP vectors must be generated.

Be careful not to step too far outside the bounds of the LP—remember it is only a linear approximation of a nonlinear and very complex set of equations. When the optimum crude mix has been determined and the crude is in the tanks or pipeline, individual units must be optimized on a continuous basis. To achieve this goal, individual unit models must be in place and a set of pseudointermediate stream prices must be developed. These models allow continuous updating of the optimum operation—that's important during frequent feed quality changes. For example, a good fluid catalytic cracking unit (FCCU) model can, on a stand-alone basis, optimize severity by trading off reactor temperature, cat-to-oil ratio catalyst activity (catalyst addition rate), and possibly feed if intermediate tankage exists or imported gas oil is run. Putting together models of highly dependent units is the next step; for example, a cat feed resid desulfurization unit and an FCCU model to optimize both units. Ultimately, having built confidence in the models, they would be on closed-loop control.

Asset teams

Organizing around cross-functional teams is an effective way to break down the functional barriers which stand in the way of high performance. Tensions and finger-pointing between functional groups are common occurrences in traditional refinery organizations ("You broke it " vs. "you didn't build it right "). Grouping operators, maintenance technicians, engi-

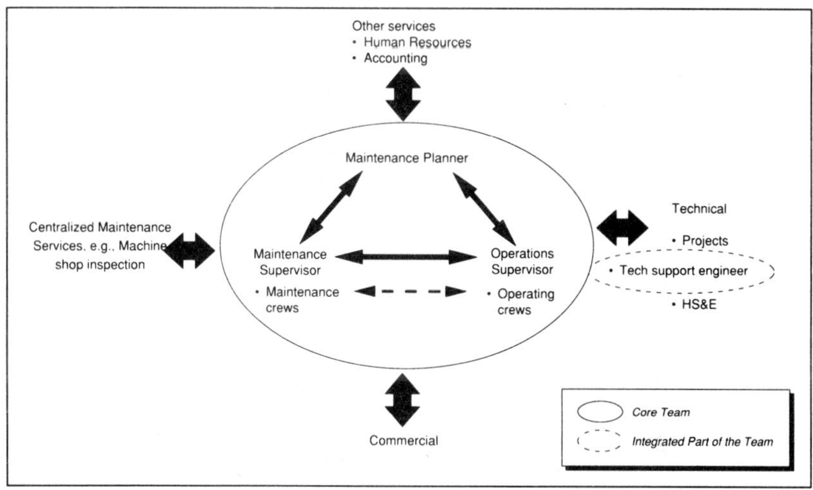

Fig. 4–9 Manufacturing Team Approach.

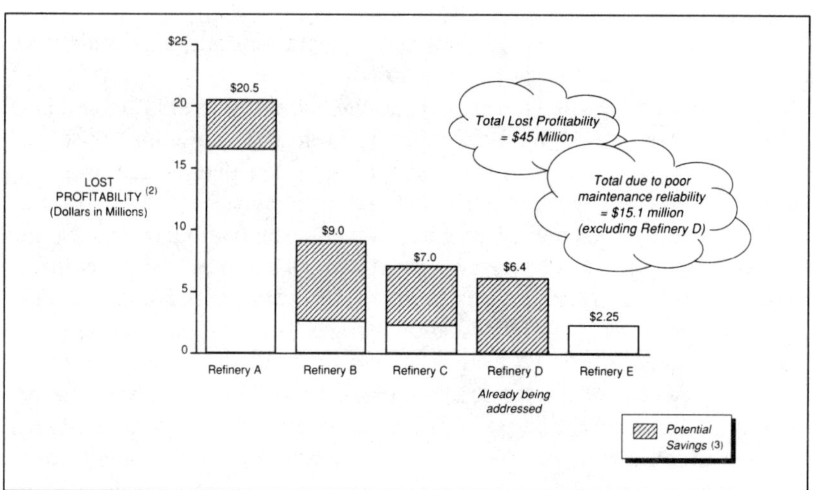

Fig. 4–10 Potential Savings from Deployment of Asset Teams. Sources: Refinery data; BA&H analysis.

neers, and a lean group of support people into a single business area, or asset team, is a powerful tool. The team structure encourages cross-functional cooperation and facilitates the establishment of economic performance measures tied to the actual performance of a given process unit or group of units (see Fig. 4–9).

Each team should be colocated and provided with the necessary budget authority and information to do their job. At the same time it is important for senior management to establish clear performance measures for each team and hold the teams accountable.

A typical refinery might have teams organized around the major units (e.g., offsites, utilities, marine terminal, FCC, coker, etc.) Figure 4–10 illustrates the potential improvement for one refinery by switching to an asset team structure. Figure 4–11 shows the benefits which can be expected from improved maintenance which should result from the deployment of asset teams.

Some additional points to keep in mind:

- Ensure that the right processes are in place to avoid optimizing an individual operating area at the expense of the overall refinery profitability.
- Have an effective process to share resources and redirect activities to the highest priority.
- Ensure that state-of-the-art consistent practices

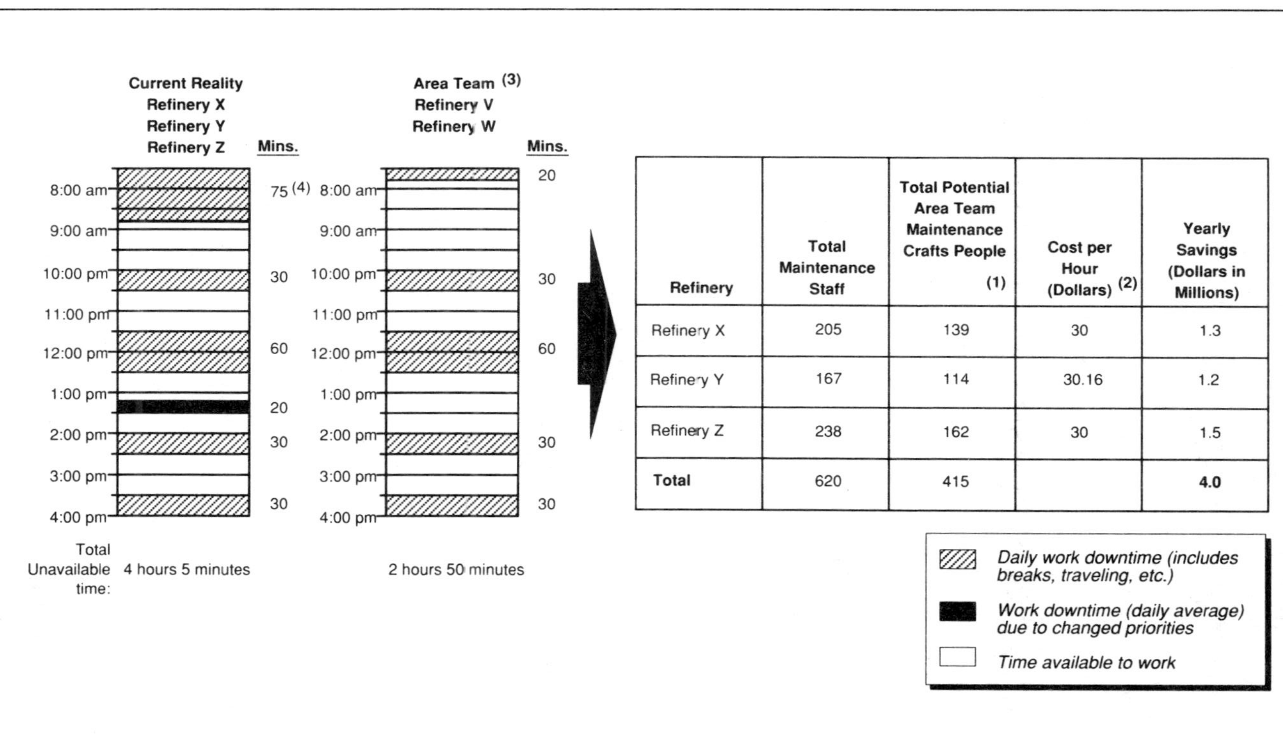

Refinery	Total Maintenance Staff	Total Potential Area Team Maintenance Crafts People (1)	Cost per Hour (Dollars) (2)	Yearly Savings (Dollars in Millions)
Refinery X	205	139	30	1.3
Refinery Y	167	114	30.16	1.2
Refinery Z	238	162	30	1.5
Total	620	415		**4.0**

Fig. 4–11 Maintenance Productivity Improvement Potential. Sources: Refinery interviews, internal reports, BA&H analysis.

are used throughout the refinery, particularly in maintenance and process engineering. A way to maintain some functional unity and career development is important.

Trust the operators

Most oil companies have in recent years delegated far greater operating authority to the refinery manager. However, very often the spirit of delegation stops in his office. Refineries are often the last remnants of oil's feudal era, because those managers impose rigid supervisory controls over the rest of the organization.

Tight controls tend to slow decision making, to discourage innovation and risk taking, and to increase costs. They also tend to hurt morale and demotivate front-line employees. On the other hand, delegation of authority downwards tends to improve morale, to boost productivity, to speed up decision-making and to encourage innovation.

Operators today are better educated and more capable of assuming responsibility than ever before. Properly trained and equipped with the right data and analytic tools, operators can usually run units closer to their optimum, thereby contributing substantially to the bottom line.

Flexibility

Most studies show that refinery operators and maintenance staff are seriously underused. These show that operators are idle almost 50% of the time. Tapping this unharnessed potential can be another powerful tool in your arsenal. Operators can efficiently and effectively perform a number of

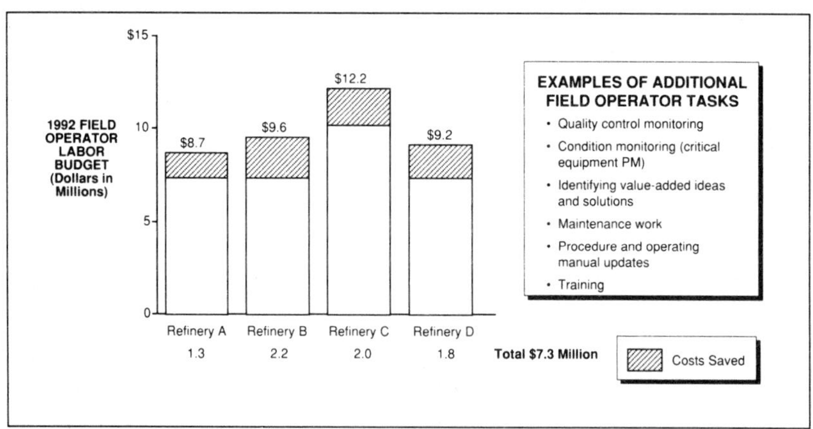

Fig. 4–12 Potential Savings from Operator Flexibility. Sources: Interviews; BA&H analysis.

routine maintenance tasks (for example, changing filters), condition monitoring, training, manual and procedure updating, project reviews, and some other support functions. Leveraging the underused time is virtually free and is bound to result in higher productivity (see Fig. 4–12).

All Maintenance Is Not Equal

The development of a differentiated maintenance strategy can have a significant impact on refinery maintenance costs. The development of full life cycle cost analysis, root cause analysis, and various predictive maintenance tools (vibration analysis, for example) is essential in today's competitive refinery marketplace. Often recurring problems are not identified unless crafts people are dedicated to teams and/or a good maintenance tracking system is in place that proactively alerts a maintenance engineer to a particular ongoing problem. See Chapter 11 for a more complete discussion of maintenance issues.

Condition-based monitoring can significantly reduce costs. Replacing equipment just because someone said it should be done every 12 months is a thing of the past. Just monitoring fouling on exchanger banks can avoid cleaning , thus significantly reducing costs and downtime, particularly during shut-downs.

Effective maintenance planning can significantly improve productivity. Knowing what needs to be done next week—at the very least, tomorrow—allows the right priorities to be determined and an efficient schedule of craftspeople's time to be developed. It then allows material to be available and delivered to the unit on time and job preparation and permits to be ready prior to the start of the craftperson's day.

Excessive emergency maintenance can destroy productivity and dramatically increase cost. It usually occurs as a result of several factors: reliability is poor; planning is neglected; poor prioritization is done; operators "game the system" to get their own particular irritations fixed; operators are inflexible and take little ownership for the assets. Addressing the causes will lower emergency maintenance, reduce cost, and increase reliability. The added attraction is that it makes everyone's day much more enjoyable. Few people relish the idea of "fighting fires" every day.

Energy is not free

Most refineries monitor energy consumption at the refinery gate; however, many do not do as thorough a job of measuring consumption by individual units within the refineries themselves. This is especially true of steam use. Yet without this metering data, it is difficult to measure the true economics of each unit. Energy is a significant portion of the operating cost, and therefore, energy management is a critical factor that should receive a high priority from every refinery manager on a daily basis.

Project Discipline

Refinery turnarounds and major capital projects eat up a lot of valuable capital. They also impose heavy burdens on refinery organizations. And the downtime required for a major turnaround, and for some major capital projects, impose additional costs in the form of lost revenues.

Project discipline is therefore an essential characteristic of a high-performance refinery. However, we find many refineries lack such discipline. The first discipline is to ensure that the project itself is economically justified. Many refineries have project engineering staffs which dream up an endless stream of work to perpetuate their existence. Often these projects will not meet the company's hurdle rate. Nevertheless, full engineering workups are frequently performed prior to conducting the economic evaluation. The solution lies in establishing an economic hurdle test early in the process to ensure that the project has a reasonable chance of economic viability.

Second, the right people must be involved up front during the feasibility study and before the design stage. Maintenance, safety, and environmental disciplines, for example, must provide input early in order to avoid costly delays and redesigns later in the project.

The third discipline is in planning and procurement. Too often major capital projects and major turnarounds are viewed as being outside the normal procurement cycle. Sometimes this is due to the heavy reliance on outside contractors, sometimes to historical accident. However, most companies have spent a lot of effort recently in streamlining their procurement operations and moving to a strategic sourcing approach (see Chapter 11 for further details). Procurement for major capital projects and turnarounds should therefore be conducted in the context of the company overall efforts and not viewed as a "one-off" event. Also, the benefits of standardization of key consumables (bolts, piping, conduit, etc.) is often overlooked.

Fourth, discipline in execution is also important. We have seen multirefinery organizations in which the time required for similar turnarounds varies enormously from one plant to another. The differences lie in the careful planning and execution of the turnaround. Far too often, there is no learning either within or across refineries of best practices (see Fig. 4–13).

In addition to these general principles, there are a number of function-specific issues which require special attention. These include:

- ***Procurement.*** We have already mentioned procurement in the discussion of project discipline above. However, this is often a major area for

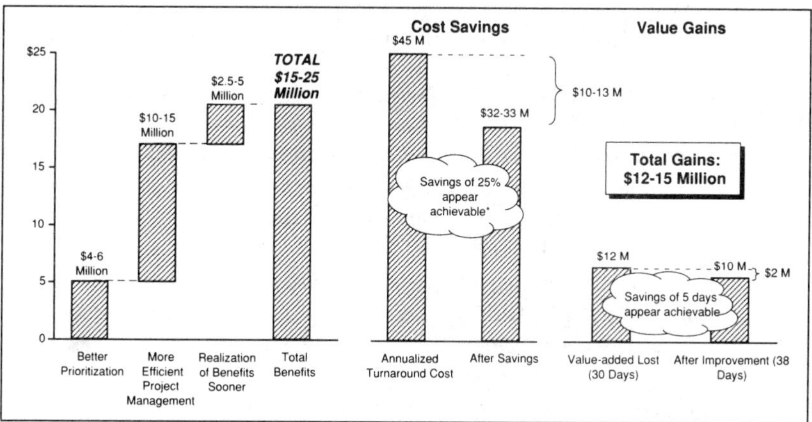

Fig. 4–13 Potential Value Gains from Improving the Project Process/Turnaround Improvement Potential.

operational improvement as well. Consolidation and standardization combined with implementation of full life cycle cost analysis can yield major benefits.

■ ***Health, Safety, and Environment.*** Refineries are under intensive scrutiny by both company and government HSE personnel. The key to success in this sensitive area is to understand the appropriate distribution of roles and responsibilities between the corporate center and the refinery. In general, it pays to place accountability with the asset team, with clearly defined performance measures monitored by a small corporate HSE group.

■ ***Process Controls and Information Technology.*** Refineries are often burdened by islands of technology which do not communicate with each other. The greatest disconnect is usually between the operational/process control systems and the financial/management systems. Moreover, many refineries have developed in-house LP models and other systems that can be better performed today by third party software like the Bechtel PIMS model.

However, in-house systems shops often view such "off the shelf" products as threatening their own empires and convince management that only a proprietary (i.e., in-house system) will do the job. Yet in-house solutions typically cost more and become outdated more rapidly than third-party products that are forced to innovate to remain competitive.

Advanced process controls offer a means to automate operations, reducing cost, improving reliability and optimizing performance. The lack of capital has often prevented refiners from making investments in this area which offers very high and short-term paybacks. Dynamic matrix control and even simple feed-forward control loops are impossible to implement on old pneumatic/local instrumentation.

HOW TO RUN A PORTFOLIO OF REFINERIES

Many of the key success factors mentioned previously in the context of running an individual refinery are equally valid guidelines for running a network of refineries. However, it is important to stress several additional points.

First, management of a refining network offers significant economies of scale for certain support functions such as procurement, technical services, supply, training, HR, and HSE.

Second, network management will usually play the major role in the capital budgeting process and in other corporate-wide initiatives.

Third, network management can identify and facilitate the transfer of best practices across refineries.

Fourth, common performance standards can be set for all the refineries.

Fifth and often, one of the largest potential impacts on the bottom line is the ability to swap intermediates between refineries in close proximity. This allows flexibility in operation particularly during unit turnarounds. The two refineries essentially approach the operation of a dual-train refinery. Of course, refineries do not have to be owned by the same company to benefit from this exchange of intermediates.

CONCLUSION

The key to economic success in refining lies in the manager's ability to meet and reconcile the economic and these operational challenges. Too often, as one client put it, refinery managers would "rather watch molecules vibrate than dollars accumulate."

---

[1] In refinery economic analysis, it is common to differentiate between "gross margin," "net margin," and "variable margin." Gross refers to the difference between the cost of the crude input and the product output. Net equals the gross margin less both taxed and variable operating costs, while variable margin refers to the net margin on incremental barrels of crude oil (it assumes the fixed costs are covered by preexisting crude runs).

CHAPTER 5

SUPPLY TRADING, AND RISK MANAGEMENT

Significant volatility has become a way of life in the oil markets. The deintegration of the supply chain from the wellhead to the street into multiple highly competitive spot markets creates the opportunity for significant volatility. Uncertain market fundamentals—politics, weather, technology—combine to make crude oil, refined products, and natural gas among the worlds most volatile commodities. This level volatility requires oil companies to develop sophisticated new supply chain management and risk management capabilities to remain competitive. Oil companies immediately began to develop supply and trading organizations, but old paradigms, built from years of integration, constrained their effectiveness, allowing new market entrants to develop powerful niches. Only now, some 25 years after the emergence of significant volatility are major oil companies developing world class supply, logistics, and trading capabilities, recapturing the initiative from the traders. This chapter explores the risks oil companies face along the supply chain and defines market driving responses to these challenges.

THE PRICE OF VOLATILITY

rude oil, refined products, and natural gas prices are exceptionally volatile—price moves of 10% and more during the course of a month are not uncommon. The differential between crude and refined products, the so-called crack spread, has proven even more volatile, ranging between \$2 and \$5/bbl during the course of a year. The marketing margin from the refined products spot market to the street has also been uncertain. Volatile quality differentials among crudes, product price differentials among grades of gasoline and between gasoline and diesel, and price differentials between and among trading centers compound the complexity of volatility in the energy markets today.

Traders find market uncertainly attractive. Traders profit from market making—earning a nice margin on the bid/ask spread on every trade in a volatile market. Traders also profit from buying and selling volatility. Traders have invested in the information necessary to price risk efficiently, buying risk when the price is low and selling risk when the price is high.

In contrast, managers find the uncertainty inherent in this environment both unsettling and expensive. Volatility creates significant economic risks, managerial risks, and capabilities risk.

Price volatility makes it very difficult to plan and invest for the long term. Upstream companies must now evaluate price volatility along with geologic risks and political risks in making exploration, development, and production decisions. Refiners have found that they must look at investments in upgrading capacity (e.g., the decision to build a new coker) as an option on the volatile price difference between residual fuel and diesel fuel. Companies with large working capital inventory positions have found that price volatility can create unpleasant, unexpected changes in financial position. Volatility makes investments in energy assets more risky and makes planning for energy companies more complicated than the tradition of net present value analysis. Oil companies must develop the capability to evaluate and to manage these economic risks to maximize shareholder value.

Volatile prices can also make managers excessively risk averse and make it harder to observe managerial performance in energy companies. One of the primary risks of decentralized decision making is that operating division managers may be unwilling to take appropriate risks (e.g., buying a whole field rather than a partial interest) because of market uncertainty. Likewise, managers may purchase far too much insurance to protect against career-limiting risk when the firm is already internally hedged. While the firm is sufficiently diversified that it should be risk taking, the manager is not diversified and risk averse. As a result, major oil companies may take on inadequate levels of risk in the marketplace, impeding shareholder returns. Many managers we talk to long for the comfortable, low-risk environment from the era of vertical integration. Corporate management must develop risk shifting and risk management mechanisms to address managerial risk to align corporate risk preference profiles with managerial propensity to accept risk.

At the same time, price volatility provides managers with convenient excuses for poor performance, making it difficult to hold them accountable. Too often, managers of profitable divisions are allowed to relax, even when the positive returns are largely attributable to market forces. The challenge of separating managerial performance from price performance in operating divisions for major oil companies presents a major challenge that few have adequately addressed to date.

Uncertainty also has a way of reshaping the competitive structure at various stages along the supply chain. Risk management approaches have radically reshaped the diesel fuel marketing system in the United States. Today, the vast majority of #2 fuel is sold under long-term or fixed-price contracts, often through traders. Those companies that have developed the ability to market these price risk-managed contracts (often through telemarketing) and those companies that have learned to price these instruments effectively, have succeeded. They have captured significant market share from the oil company majors who initially viewed this type of marketing as aberrant speculation. Volatility creates a new market in risk—those companies able and willing to buy and sell risk are able to gain a significant competitive advantage over those companies who only want to sell products.

Deintegration and market uncertainty have reshaped competition in import/export markets around the world. Crude and refined product flows now shift with the tradewinds created by the world oil markets in Rotterdam, Singapore, New York, and Houston. Sourcing patterns continually shift as traders move to eliminate arbitrages across markets. Yet even today, many oil companies have a tendency to move their own molecules (crude or products) to their own markets or to hardwire sourcing patterns into their supply plans. Major oil companies who traditionally served the Boston market out of Houston continued these patterns long after new competitors entered these markets and began to win market share by importing product from Canada, the Caribbean, and Northwest Europe at substantially lower cost. Uncertainty places a premium on flexibility, speed, and agility—characteristics which oil company majors did not require when vertical integration was the norm.

Regrettably, most companies assume away economic risk, ignore managerial risk, and defer decisions about capability risks until it is too late. Most companies assume that they are well diversified and well hedged without doing a careful risk analysis. Oil companies often spend more time managing foreign exchange risk than they do managing oil price risk. Our analysis suggests that even vertical integration among production, refining, and marketing rarely provides a strong financial hedge even within a market. Likewise, basis risk across markets creates new risks in oil company portfolios. Supply and trading organizations must develop the capability to apply risk management tools and techniques to the fundamental corporate strategic decisions in today's volatile environment.

Many companies respond to concerns about managerial risk with a faint admonition to managers to ignore performance measures when the measures conflict with corporate interest economics. In corporate cultures that place a strong emphasis on performance, managers typically ignore the words and watch the performance numbers and avoid risk assiduously. The best competitors have developed approaches that use the supply function as a central clearing house for risk, providing managerial risk insurance efficiently, aligning managerial interests with corporate interests, and enabling line management to focus on taking appropriate risks for the company.

Similarly, most large oil companies have been slow to develop the capabilities necessary to manage in a highly volatile price environment. Tightly integrated supply chain management business processes, rapid new risk management product innovations, and performance measures that facilitate and reward flexibility have only emerged slowly in the oil industry.

TRADING IN RESPONSE TO UNCERTAINTY

e believe energy companies should take on new supply roles, including risk management, gradually adopting increasingly sophisticated roles over time (see Fig. 5–1).

Currently, most oil companies have developed sound portfolio hedging strategies for crude oil, but retain transaction hedging paradigms for refined products. Our surveys of oil companies make it clear that a sound supply and risk management strategy allows companies across the supply chain to take prudent operational risks instead of assuming that volatility was unmanageable. Companies with sound supply and risk management strategies have been leaders in offering fixed price contracts, developing highly profitable inventory build/draw strategies, designing optimal spot-term portfolios, and introducing other important innovations in supply strategy.

However, taking shortcuts can be costly. Successful strategies must be built on a firm foundation of analysis and practice, as illustrated in Figure 5–2.

Sound risk management strategies must begin with a clear agreement on what is at risk. Risk management must have a clear baseline. Most companies view spot supply to be at risk while they ignore the risk inherent in fixed price term supply contracts, in-transit inventories, inventories in storage, and fixed-price sales agreements to their customers. The best supply and risk management programs are built on a very clear understanding of what creates risks and what risks the firm is willing to take.

Efficient supply and risk management strategies are built on a clear assessment of net risk. The best companies look for internal hedges before they turn to the outside market. Measuring a firm's net exposure allows a company to separate economic from managerial risks. Firms can hold managers risk neutral without using risk management where self-hedges exist. The size and the shape of the firm's net exposure defines the company's economic risk and indicates the level of risk management which may be required. Based on the characteristics of this net exposure, companies can choose risk management tools—futures, options, swaps, caps, collars, floors, swaptions—appropriate for that company's risk profile. The characteristics of the company's net exposure can also help the company think about supply organizational structures: central trading desk, supply business unit, supply profit center.

For those companies who decide to trade more aggressively, trading a "practice book" (trading on paper only, not in real adjustments) provides

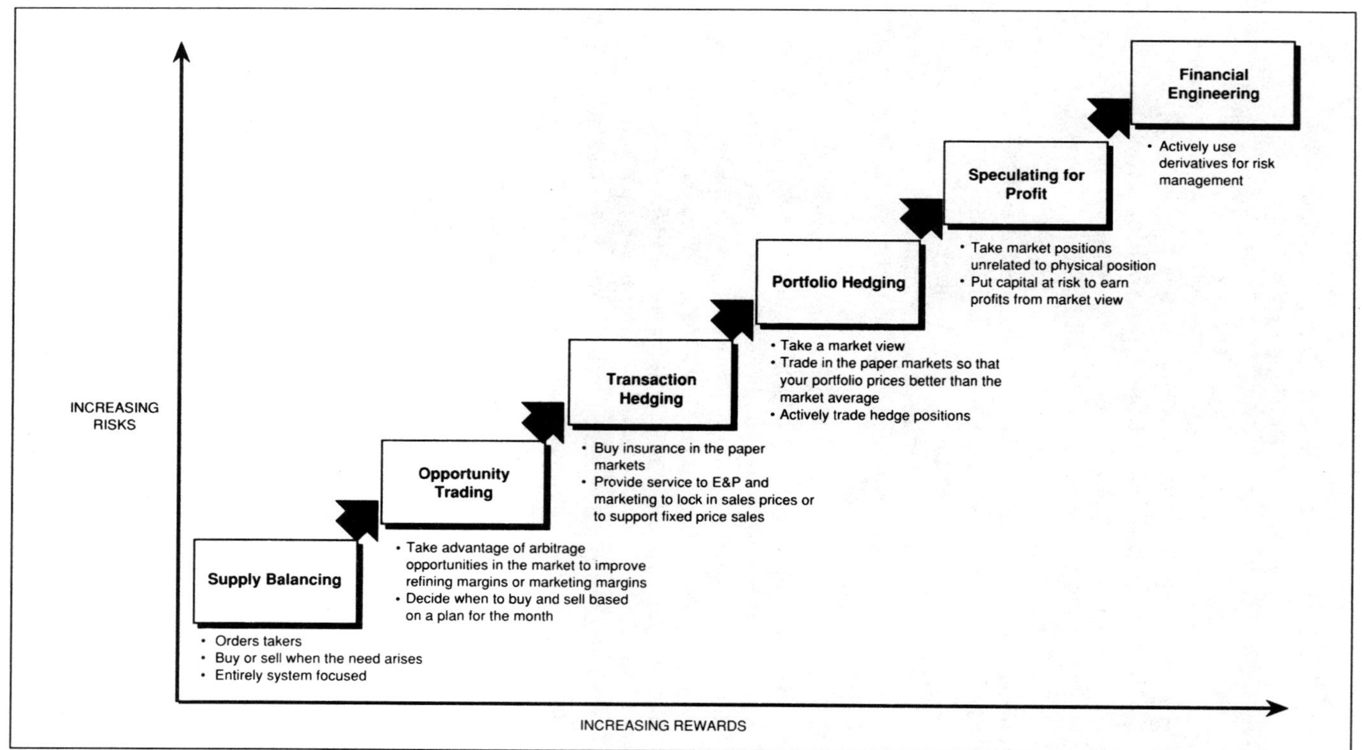

Fig. 5–1 The Spectrum of Roles Played by Supply.

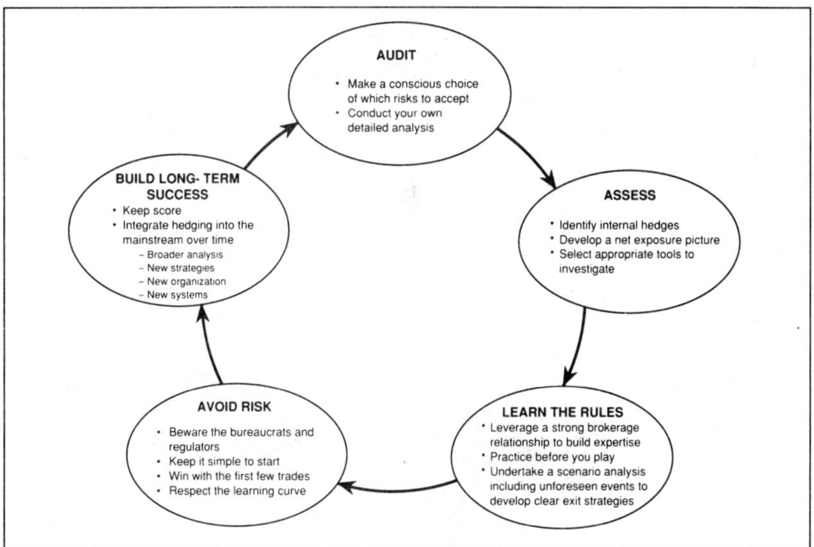

Fig. 5–2 Principals for Beginning a Risk Management Program.

valuable experience and tangible evidence of success to support going forward. The best supply and risk management strategies incorporate detailed scenario analyses, focused particularly on exit strategies. These strategies examine the distribution of returns across a range of scenarios and plan contingencies to deal with potential high risk situations.

Effective supply and risk management programs assiduously avoid risk:

■ ***Bureaucratic Risk.*** If senior management does not understand and support the program, the argument against undertaking risky supply strategies may be overwhelming. All too often, managers set up uneven treatment for risk and reward and severely second-guess trading strategies based on hindsight. Bureaucracy can cripple sound supply programs.

■ ***Complexity Risk.*** The best companies follow simple strategies to start. The efficiency of a complex strategy may be appealing, but these benefits can dissipate quickly if the market moves in unexpected directions of if there are unforeseen cash flow complications.

■ ***First Trade Jitters Risk.*** Companies that have begun trading find that management feels better if the first few paper positions make money. Even though this means that the physical positions have lost money, the managers like to know that the insurance they are buying actually pays out.

■ ***Learning Curve Risk.*** Very strong traders estimate that it costs perhaps a million dollars in opportunity costs to develop a good trader. Hence, the best companies allow trading staff to focus on supply and risk management, making a career out of trading.

In time, companies that build effective supply and hedging programs progress into portfolio hedging. These companies calculate the net exposure of the companies whole supply portfolio. They adjust risk assessment to account for demand patterns. They separate risk management from basis trading. They actively manage hedge positions against the entire portfolio. They begin employing sophisticated trading strategies, including spread trading, synthetic hedges, and building rather than buying straddles, collars, caps, and floors.

Over the longer term, companies integrate risk management into their overall strategic plan. These companies build advantaged, high-margin customer relationships around risk management services. They lock in margins on a portion of the sales portfolio while allowing another portion of the sales portfolio to fluctuate.

We believe that companies that undertake a disciplined approach to risk management can be quite successful. The key is to lay the foundation carefully and take on new roles gradually.

LOGISTICAL COMPLEXITIES

At the same time, the physical supply chain has become more complex to manage. Deintegration has introduced numerous handoffs along the physical supply chain—refined product may be sold from a refinery to a trader who moves the product using independent marine or pipeline transportation to an independent terminal operator before selling the product again to an independent marketer who moves the product on a common carrier truck to a service station owned and managed by a branded distributor. Environmental rules and regulations require new fuel formulations that vary by region, increasing dramatically the number of petroleum products supply organizations must manage.

Efficient and effective supply chain management requires viewing and controlling the petroleum supply chain as an integrated whole from the crude source to the street. Information, contracts, and business processes must substitute for physical integration today. Volatility is inherent in end use demand, driving up distribution and inventory carrying across the entire supply chain. The standard deviation of daily demand for a typical gasoline distributor exceeds 20%. Supply and demand volatility requires companies to develop strong, analytically based inventory management and scheduling processes and tools. Even today, most oil companies fall well short of best practices for managing the supply chain in a volatile environment.

INTEGRATED SUPPLY CHAIN MANAGEMENT BUSINESS PROCESSES

Volatility is inherent in the end use demand and is transferred through dealer behavior with limited benefits through pooling. Several oil companies have begun to manage station inventories to force down demand volatility and reduce distribution costs. Also, more effective rack pricing can smooth demand while increasing average price realization.

Appropriate safety stock levels can be set using forecasting and optimization models. Supply and demand volatility drive terminal inventory levels. Volatility and refinery economics drive refinery inventory levels. Companies may also benefit from managing aggregate inventory exposure to take advantage of market signals.

An integrated sales and operations planning process is essential for successful supply chain management. World class companies manage the supply chain as a critical strategic asset, developing and executing sound strategies for inventory management, product changeover, and pricing (see Fig. 5–3).

Rationalizing terminal infrastructure and revising terminal operating practices to minimize Ried Vapor Pressure (RV) changeover costs and reduce administrative overhead are essential for effective supply chain management. Upstream of the terminal, oil companies must seek improved sourcing, trading, and exchanges. Improper sourcing can make a company seasonally uncompetitive in key markets. A terminal-by-terminal analysis can identify sourcing arrangements that destroy value—and better alternatives. Exchanges need to be managed to avoid averse economics (see Fig. 5–4).

At the refinery level, sound supply chain management can improve unit utilization while reducing intermediates and crude inventory. Refinery optimization benefits from a clear future view of the market. Refinery operations can be improved by removing excess inventory which hides underlying operating flaws. Refinery crude slates can be enhanced by taking a statistical approach to the question of fixed vs. variable crude slates.

	Level 1 (Novice)	Level 2	Level 3	Level 4	Level 5 (World Class)
Hierarchical Supply Chain Control	Supply chain managed purely at an operational level	Some level of policy setting by senior management—but chain managed mainly at operational level forecast errors can impact planning	Ad hoc scheduling attempts to resolve forecast errors and supply/demand imbalances	Three integrated hierarchical levels of control—policy setting, short term planning and scheduling with different resolution and time horizons	Linkages at strategic, tactical and operational levels extend to customer and supplier
Well Defined Supply Policies	Undifferentiated supply policies—all products made to meet short-term demand Uniform levels of customer service	Selective supply policies – e.g.some seasonal build strategies Policies developed at operational and tactical levels through ad-hoc process	Supply policies articulated for products in terms of demand requirements with limited supply and transportation considerations No reconciliation between policies, inventory, service and utilization objectives	Policies developed based on economic alternate source, inventory utilization and service trade-offs Customers are given differentiated service	Supply policies articulated for end products. Intermediate products and raw materials in terms of supply lead times, reliability, batch size and supply source Policies developed at senior management level via tradeoffs
Analytically Based Inventory Targets	Inventory targets determined by financial ROCE targets Targets set at aggregate level by product group or business unit	Aggregate inventory target determined by financial targets Minimum and maximum levels of inventory developed by ad-hoc process	Inventory targets calculated statistically based on reorder point techniques Targets could be developed for high volume items only	Inventory targets calculated based on internal lead times for minimum demand variability, batch size, service levels Targets set at all product levels	Inventory targets calculated based on lead times supply and demand uncertainty, batch size, transportation costs and service levels Targets set at raw and intermediate material levels. Updated to reflect seasonality
Robust and Integrated Planning and Scheduling	Planning subsumed within operational scheduling Weekly resolution and monthly horizon	Long term capacity planning driven by max level forecasts and performed at product level Short term scheduling based on mixture of forecasts and actual orders or target resupply points	Long term aggregate capacity planning at product level Manufacturing optimization done by LP – time scheduling performed by manufacturing scheduling	Various functions are integrated in supply chain management Short term scheduling driven by "pull" systems with feedback to planning system	Proactive involvement by production, materials management and suppliers to upgrade supply chain flexibility. Cross-functional supply chain management teams led by a high level leader All manufacturing facilities optimized jointly
Time Phased Supply/Demand Planning Systems	Raw material, manufacturing and distribution schedules uncoupled and derived independently No visibility of supply and demand along the chain	All scheduling driven off a materials requirement planning system Little or no visibility of supply/demand order status and inventory along chain	Planning/scheduling performed by an integrated system Control parameters derived from operational experience, scheduling performed manually to daily resolution Limited visibility of availability along chain	Scheduling driven by pull systems integrated with order processing systems all along the chain Visibility of demand and supply along the chain is ensured	Supply chain systems enabling supply/demand balance at daily resolution all along chain Routinized and highly automated scheduling tools
Demand Planning	Best Guess Through Information Market Intelligence No match to supply capability	Statistical forecasts of mix level produced by central marketing function Accuracy rarely monitored	Statistical forecasts at mix and volume levels (medium and short term). Produced by region. Central consolidation and reasonableness check. Accuracy monitored	Supply and demand matched at a minimum weekly resolutio Methods and data updated based on past accuracy performance	Statistical forecast of mix and volume levels by location Central consolidation Modified to reflect promotional campaigns, price changes, and other competitive/macro-economic factors

Fig. 5–3 Best Practices in Inventory Management and Scheduling.

	PASSIVE	ACTIVE	SOPHISTICATED	INNOVATIVE
Sourcing	**Proprietary** • Source primarily own barrels • Term or exchange to balance	**Term/Exchange** • Source from secure supply • Spot trade for rebalancing	**Local Trading** • Source from optimum local spot • Rely on brokers to find deals	**Global Trading** • Source from optimum market worldwide • Buy and sell to adjust position
Trading	**Supply Balancing** • Wet order takers • Entirely system focused • Buy or sell when need arises	**Basis Trading** • Take advantage of buy-sell spreads • Decide when to buy and sell based on monthly plan • Buy insurance in the paper market • Combined wet and paper trading • One hour decision making	**Portfolio Hedging** • Specialized Wet Traders • Specialized Paper Traders • Clear performance measures—traders must beat market average • Trading inventory programs for all products • Fast decision making (20 minutes or less)	**Market Making** • Take market positions unrelated to supply needs • Put capital at risk • Buy and sell based solely on market view • Profit from buy-sell spread • All inventory is trading inventory
Exchanges	**Gentleman's Exchanges** • Balance out annually • Exchange for convenience only	**Geographic Exchanges** • Track barrels • Manage exchange balances to zero	**Traded Exchanges** • Manage exchange balances as spot barrels • Track dollars	**Opportunistic Exchanges** • Seasonal • Based on snow line • Strategic rack receipts only

Fig. 5–4 Best Practices in Sourcing, Trading and Exchanges.

ORGANIZING FOR SUPPLY SUCCESS

ffective supply and trading depends on new organizational structures. The traditional organization segmented into functional units is ineffective for identifying, understanding, and managing risk. Likewise, organizational silos prevent effective supply chain optimization. More horizontal organizational structures that link the customer to the fuel supply acquisition strategy have proven to be more effective in managing both the physical and financial risk exposure (see Fig. 5–5).

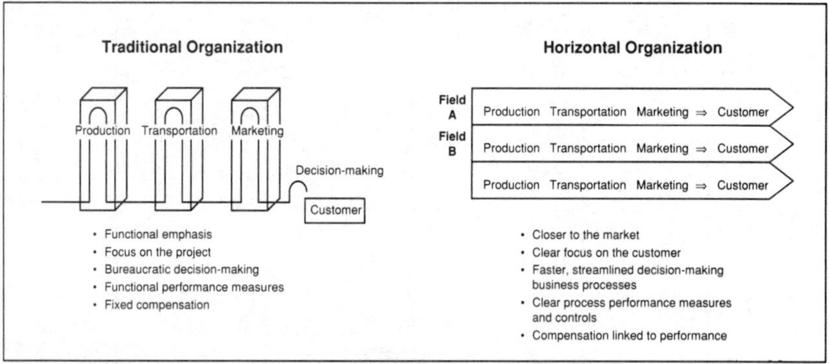

Fig. 5–5 Alternative Organizational Approaches.

Through tight horizontal integration, energy companies can reduce direct labor expense and improve operating effectiveness. Tight integration allows supply organizations to eliminate redundant positions across organizational boundaries. Tight integration also facilitates strong cross-functional decision making.

TRADING

A sound supply sourcing strategy combined with integrated trading operations can allow downstream, crude, and products trading organizations to operate with efficient trader headcount. The role that supply plays is critical for trader staffing levels as well—trading for profit requires substantially more trader headcount per barrel than supply spot trading. Also, there are substantial economies of scale in crude and products market coverage. Organizations that centralize domestic and international crude and products trading tend to be able to cover the market with fewer traders while at the same time making effective tradeoffs such as: make vs. buy products, international versus domestic. Trader-trader linkages are critical wherever optimization across the barrel or across geographies is possible.

OPTIMIZATION

The most effective supply organizations integrate crude supply planning and refinery optimization decision making within the supply organization. Traditionally in the downstream, refining, and supply have operated as separate organizations—refining has traditionally had veto power over supply decisions on crude slate, product mix, and run rate. However, some companies have found that they make better decisions with fewer staff when the refinery planners and the supply planners are combined into one group (usually located with the traders) that has the authority to optimize systemwide and across the barrel. The same model applies to the relationship between supply and marketing. This type of organization makes decisions much more quickly, accurately, and efficiently than the traditional alternative.

SCHEDULING

Likewise, scheduling organizations—pipeline, marine, rail, and truck—can benefit from tightly integrated organizational structures and decision making. All too often in major oil companies, crude and products supply schedulers nominate to pipeline and marine schedulers within the same organization. Similarly, movements which can take place on more than one mode of transportation often require multiple handoffs. This level of scheduling fragmentation increases the amount of staff required and significantly decreases decision-making effectiveness in most supply organizations. Further downstream, terminal inventory management depends both on effectively scheduling the supply of product into the terminal, but also scheduling the liftings of product out of the terminal. Tight integration of scheduling in and out of byproducts terminal can dramatically improve decision-making effectiveness and reduce the headcount required. Finally, strong trader-scheduler coordination can improve market coverage while maintaining or reducing the required headcount. In the most efficient supply organizations, schedulers are an active arm of trading, handling in-line trades, convenience exchanges, and term exchanges.

OVERHEAD

The most effective supply organizations tend to have relatively little administrative overhead: long-term planners, short-term analysts, managers and support staff. The supply organizations at most major oil companies tend to be burdened by bureaucratic paradigms that suggest that detailed, long-range planning is necessity, front office staff cannot be trusted to do back-office work, senior staff should manage people, and people with experience require secretaries. Effective supply organizations can substantially reduce their administrative overhead by breaking down these paradigms. Tightly integrated decision-making teams (e.g., refinery-supply optimization) substantially reduce the amount of paper work, traveling back and forth across organizational lines. The most effective supply organizations recognize the limits of supply planning: most long-term planning needs to be done outside of the supply organization with limited supply input—supply's short-term focus often provides little insight into long-term trends—while most short-term planning needs to be conducted at the front lines by operating staff—large planning organizations tend to impede rather than facilitate effective short-range planning.

Effective supply organizations tend to substitute systems for support staff. Good traders put their own deals into integrated financial, performance tracking and contract administration systems, rather than scribbling their directions on a piece of paper and tossing the deal over the transom. The best traders are also responsible for the quality of their transactions—asking accountants to resolve trade discrepancies is both inefficient and ineffective. The best trading organizations have very wide spans of control with many player-coaches and few "go-to-meetings" managers. In this type of environment, positions like trader-secretary, manager-secretary, and scheduler-secretary become oxymorons.

LINKAGES BEYOND SUPPLY

Integrated supply chain management can also reduce headcount between the supply organization and pricing, between terminal operations and light products pipeline operations, between E&P operations and crude pipeline operations, and between terminal operations and trucking operations. While these activities may be outside the immediate scope of this effort, the same principles apply: addressing these issues in the same way may enhance supply efficiency.

SUPPLY INFORMATION SYSTEMS

ffective supply and trading also depends on a strong information systems infrastructure. The advent of the oil futures market in the United States made information more valuable, but it also increased the speed of information flow, making it harder for managers to keep abreast of market developments (see Fig. 5–6).

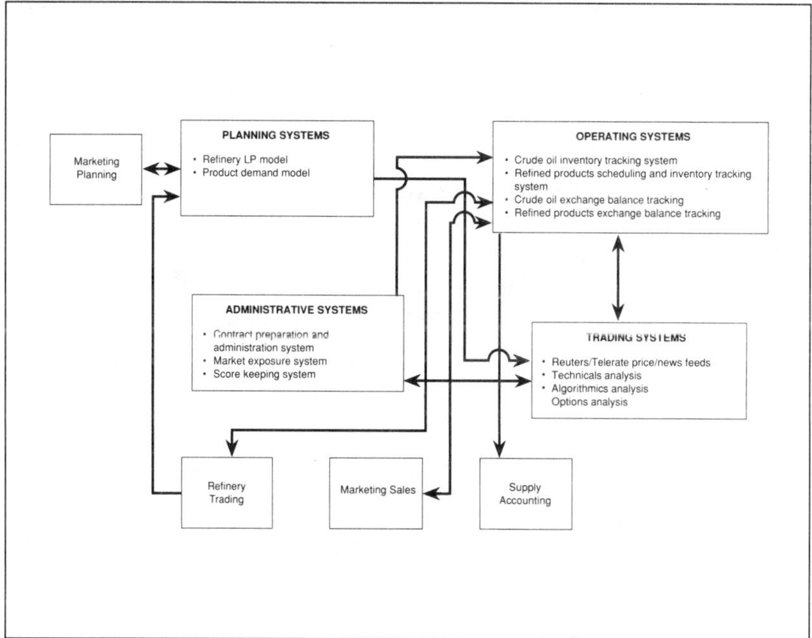

Fig. 5–6 Illustrative Systems Map.

Information systems can either become a source of competitive advantage or a costly management distraction. Strategic investments in systems technology for trading analysis, inventory management, scorekeeping, and credit analysis can pay out in less than two years. Yet all too often, we see supply and trading information systems projects coming in years late, 100% and more over budget, with outdated technology and with functionality well below the industry average. Trading managers and IT managers who propose these projects with great confidence and work nights and weekends to complete the projects find themselves feeling like Sisyphus when the new systems comes online—all that energy for naught! We believe success is both necessary and possible: technology is available that can radically reshape the way you do business, driving down costs and risks and driving up profits, and an effective management framework can deliver this technology to the frontline on time and under budget.

Strong information systems underlie effective business processes and organizational structures, providing an integrated package of information and analysis quickly to individuals throughout the organization. With strong information systems, everyone can respond to the same high-priority market signals; without these tools, strategic goals become fragmented as each part of the organization sifts through a stack of ambiguous information, responds to a different set of external drivers, and wastes time with reconciliation.

Information, speed of response, and superior execution have become major levers of competitive advantage. Information technology can provide a powerful competitive advantage across the spectrum of supply and trading activities. Imagine the following scenarios:

■ ***RVP Changeover.*** One company, using real time data fed by remote tank monitors, carefully manages its gasoline inventories, allowing it to meet costly RVP targets days later than their major competitors, while purchasing only a small amount of blendstock at fire-sale prices. Another company, relying on weekly, manual gauging reports must changeover early, buying large quantities of blendstock at the peak of the market and selling low-RVP product before it's needed.

■ ***Refinery Emergency.*** One company faced with a sudden refinery emergency—dramatically reducing heating oil output—successfully enters the market, covers the product deficit, and takes a long, speculative position before the market reacts. Another company's supply group finds out about its own refinery outage from the competitor whose facility is next door, takes several days to determine how to unwind the exposure, and pays a hefty fee to cover.

■ ***Trading in the Pipe.*** One company, hearing reports that a competitor is out at a key marketing terminal, quickly checks its inventory levels and sells incoming product to the distressed competitor at a premium in the line, secure in the knowledge that replacement inventory will cover just in time. Another competitor watches the trade take place, unsure where its inventory levels stand.

■ ***Credit Risk.*** One company hearing of the assassination of the prime minister of a major refined products importing nation, performs a credit analysis within a half hour of the report and adjusts its risk management strategy to reduce its exposure. Another company takes weeks to determine what it is owed when a major trading partner goes bankrupt.

■ ***Hedging.*** One company takes advantage of a sudden cold snap to sell off heating oil inventories, while hedging replacement barrels with futures and swaps. Another company, uncertain of its exact inventory, builds for security in the contango market, selling the excess inventory off at a loss in the spring.

■ ***Derivatives.*** A company with a sophisticated derivative valuation software offers a range of new pricing options to its customers and prices the required risk management efficiently; another company loses share as it keeps to its traditional offerings.

An information systems strategy focused on the levers of strategic advantage must underlie systems acquisition decisions. Based on our work with major supply and trading organizations, it is clear to us that trading, pricing, and logistics systems offer the most immediate rewards. Price feeds, news feeds, trading alarms and triggers, portfolio management tools, and desktop charting are now required equipment for a successful trading operation. Online score-keeping systems (including wet, paper, and logistics costs), online contract management (including deal approvals), and online financial accounting can also provide an advantage in efficient trading operations. The most aggressive traders are experimenting with desktop synthetics, fuzzy logic devices that codify the unconscious skill of talented traders and neural networks.

Logistical systems can be used to reduce inventory levels 20%–30% and more. Real-time actual inventories, sales, arrivals, exchange agreement tracking, scheduling optimization programs, and forecasting, scenario analysis, and dynamic rebalancing tools each contribute to optimizing the logistical equation. Similarly, sales and pricing systems—rack pricing databases and analytics with warning flags, sales tracking, trending, forecasting, and price deployment systems that announce changes to the market at the last moment possible—can make a substantial contribution on the revenue side.

However, the best strategy in the world will fail if systems structure issues are ignored. Database and data structure decisions influence the effectiveness of systems investments dramatically. Even then, a strong business case must be made for new software and hardware investments based on increases in trading values, improved refining values, decreases in back-office processing costs, and/or lower systems development and support costs. Those companies with the greatest success have re-engineered their business processes at the same time they adopted new systems.

When the rate of technology change is combined with increasing market efficiency and shifting market psychology, the value of speed and flexibility rises again. Each new generation of trading systems raises the ante as better information tightens spreads further. Failure to keep up can make trading a less than break-even proposition. Meanwhile, shifting market psychology, and the proliferation of new derivative products places a premium on different pieces of information, or a different type of analysis every several months—last year it was candlesticks; next year maybe the bakers will have their day. This competitive spiral requires continual upgrading of trading systems while at the same time narrowing the revenue stream that pays for new systems—doing things better rather than doing things faster is the only way to make systems pay in this environment.

Information systems can be used to strengthen the bonds among strategy, business processes, and organizational structure within major energy companies. Information can be a source of competitive advantage, driving down inventory levels, speeding crude substitution decision making, enhancing make-buy decision making, making hedging more precise, and facilitating emergency response and in-pipe trading.

Information systems can reinforce the importance and power of supply management teams. Systems can be used to segment and route market signals to the appropriate teams (e.g., looking at a particular pipeline rather than a geographic region). Scorekeeping systems can support significant increases in delegations of authority—managers can monitor risk while allowing the frontline troops to make market-responsive decisions. Global information and communication systems can increase the opportunities to find and capture high-margin opportunities. Refinery-supply and marketing-supply linkages can integrate market data and operating data making operating decision making more market-responsive and reducing the impact of functional chimneys.

The challenges of implementation are profound, but the rewards can be dramatic. A strategic approach to systems development can ensure that information systems become a source of competitive advantage, allowing trading managers and technology managers to avoid the curse of a costly distraction.

LOOKING TO THE FUTURE

e expect market risk to increase in the future. The future will see the emergence of a multitude of smaller, oil-linked financial markets. The retail market for financial risk management will continue to increase at a rapid pace. The oil-linked financial markets will also become more global. We expect that the volume of swaps transactions will exceed the volume of futures transactions as more and more companies seek tailored, off-balance-sheet financing and risk management alternatives.

These future trends bring with them significant new risks. Swaps substitute credit risk for absolute price risk and basis risk. The proliferation of sophisticated derivative instruments creates a significant complexity risk in the marketplace. These trends in the world petroleum markets create significant financial evaluation risk for traders and shareholders. The physical distribution network will continue to become more complex. The introduction of reformulated gasoline will fragment the distribution system. At the same time, terminal rationalization activities are reducing schedulers' degrees of freedom.

To remain competitive, energy companies must continue to innovate along the supply chain. Companies must adopt ever more sophisticated trading and risk management practices to shape and control risk exposure. Likewise, superior supply chain management business processes can become a strong source of competitive advantage from the crude source to the street.

---

1 This chapter draws on previous articles published in *The Energy Journal, Futures, Natural Gas Week,* and the *Oil & Gas Journal* and speeches to seminars hosted by Executive Enterprises and Sun Microsystems.

CHAPTER 6

GASOLINE RETAILING

Major oil companies are engaged in a diverse and complex set of activities that range from the exploration for raw materials through the manufacture and marketing of finished products. However, their public profiles and brand images are concentrated primarily in just one of these activities: gasoline retailing. In the minds of most consumers, the corner service station *is* the oil business and the people who work there are its employees.

The gasoline service station is indeed the dominant distribution channel for oil company products, and it is the focus of this chapter. Gasoline is one of the most important retail businesses by almost any measure. The average American adult visits a gasoline station at least once per week, and the retail value of petroleum products sold in service stations ranks as a major component of most household budgets. Gasoline stations occupy the busiest intersections and the most expensive real estate.

This chapter provides an overview of gasoline retailing business roles played by major oil companies. It summarizes the physical flow of product and the structure of asset ownership. While not overly complex, the industry structure is more sophisticated than most consumers perceive. In addition, we will discuss the evolution of retailing formats and services, the economics of the service station, and the role of marketing and brand management in gasoline retailing. While the discussion draws heavily on North American examples, it is generally relevant to retail gasoline marketing throughout the world.

THE SUPPLY CHAIN

his section describes the movement of product from the terminal to the service station, where it is ready for retail sale. Stations generally have an underground storage tank to hold each grade of gasoline: regular, intermediate, and premium. These underground tanks feed the dispensers used for filling auto fuel tanks. At least one company, Sun Oil, has developed a system for blending at the pump". As a result, Sun's stations can have two tanks rather than three, one holding regular and the other premium. At the pump the motorist chooses the octane he wants, and the requested volumes of regular and premium are mixed during fueling (see Fig. 6–1).

The unit cost of delivery is strongly affected by two variables. First, the further the distance from the terminal to the station, the greater the period of time required to complete delivery. Extra time adds labor cost and effectively lowers asset usage (i.e., a terminal with widely scattered stations requires more delivery trucks than a terminal with nearby stations).

The second major variable is delivery efficiency. It is far cheaper, on a unit basis, to deliver whole loads rather than partial loads. Trucks leaving or returning half-full are a sure sign of an inefficient delivery process.

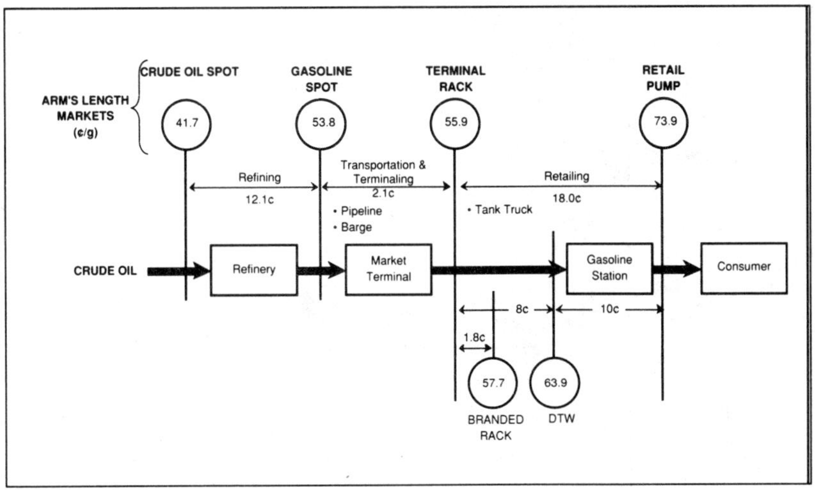

Fig. 6–1 Gasoline Retailing Supply Chain.

The price of gasoline depends on its location in the supply chain from terminal to station. Unbranded gasoline is sold at the terminal at a price that generally tracks refinery spot market prices plus the pipeline or barge transportation cost from the refinery to the terminal. In other words, there is very little value added in selling unbranded gasoline above the product cost, storage, and accounting and contract administration costs. Unbranded gasoline is sold as a commodity and its price is very competitive, closely tracking the value of spot gasoline at the refinery plus transportation and overhead costs (see Fig. 6–2).

Gasoline is branded by blending a set of additives at the terminal rack. The exact composition of the additive package is proprietary for each company, although there also are available generic additive packages for gasoline. Typical branding costs are roughly 0.5¢–1.5¢/gal, as shown in Figure 6–3. Administrative and overhead costs are addition to the physical cost of the additive package. Oil companies usually charge 0.5¢–1.5¢/gal more for branded gasoline than for unbranded regular gasoline. This premium is intended to cover all branding costs and to capture the incremental value of the brand.

In practice, this also is an extremely competitive business, with gross sales margins of around 2%–4% for regular unleaded gasoline. Much of this gross margin is offset by direct branding costs and by related costs such as advertising. On the other hand, for premium grade gasoline, the unbranded-branded spread is typically on the order of 2¢–3¢/gal. In short, most major companies make their money on sales of premium.

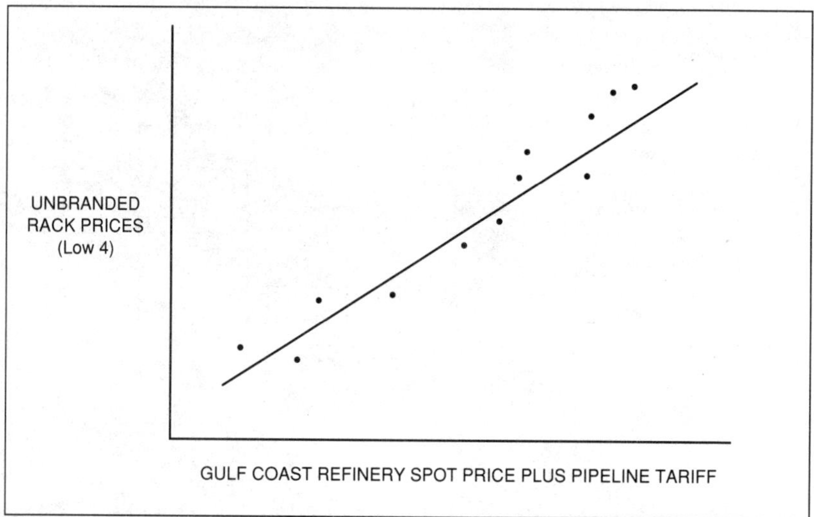

Fig. 6–2 Relationship Between Unbranded Rack Price and Spot Prices.

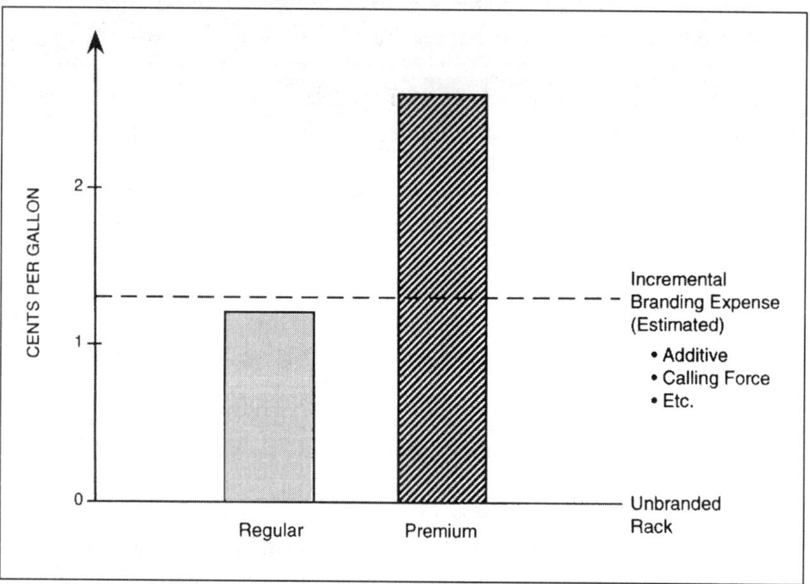

Fig. 6–3 Branded Rack Uplift Over Unbranded Rack.

Just as the spread between the cost of gasoline delivered to the terminal and unbranded rack has to cover the cost of operating the terminal facility, the spread between the branded rack and retail price has to cover the cost of trucking and station operation. This margin on a national basis was roughly 20¢/gal in 1993. The actual delivery cost for gasoline obviously varies with distance from the terminal to the station, but for a typical network of a terminal and stations it is roughly 1¢–1.5¢/gal. The remainder of the "rack to retail" spread covers the cost of the land and improvements for the service station as well as its ongoing operation and physical maintenance.

The spread between terminal and station varies both over time and across geographies. A number of factors impact the spread, for example:

- Rising or falling crude oil prices
- Seasonal demand
- Competitive pricing moves by existing companies
- Entry or exit by companies from the market
- Requirements for oxygenated gasoline

Similarly, the rack to retail margin varies substantially by market. These variations are driven primarily by competitive dynamics among companies.

CONSUMERS

asoline is often viewed as the ultimate "grudge purchase." The typical car owner spends at least $500 per year on gasoline at current prices and needs to consider the cost of gasoline as a major personal budget item. At the same time, motorists never see gasoline, yet they have to stop in heat or cold to purchase it, and they usually pump it themselves. These are less-than-ideal selling circumstances. Nevertheless, consumers have definite attitudes toward gasoline that marketers can use in merchandising their product.

Because gasoline purchases are often forced rather than planned, consumer buying behavior is somewhat erratic. A motorist's prejudices have to be resolved against the local service station situation when the fuel tank is low. As a result, people who normally buy a brand or grade of gasoline alter behavior for the sake of convenience (see Fig. 6–4).

There are relatively strong consumer loyalties to locations. The location loyalty is likely borne out of convenience: a particularly convenient

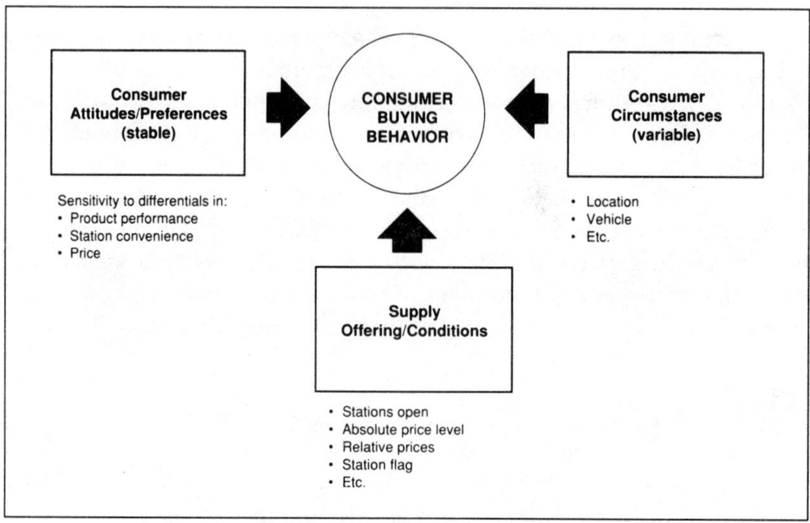

Fig. 6–4 Principal Influences on Consumer Gasoline Purchasing Behavior.

location for a segment of the local population is likely to attract business. There also is stability to this convenience, since new station volumes tend to stabilize within a year of opening and often remain that way until competitive dynamics change, such as another station open nearby (see Fig. 6–5).

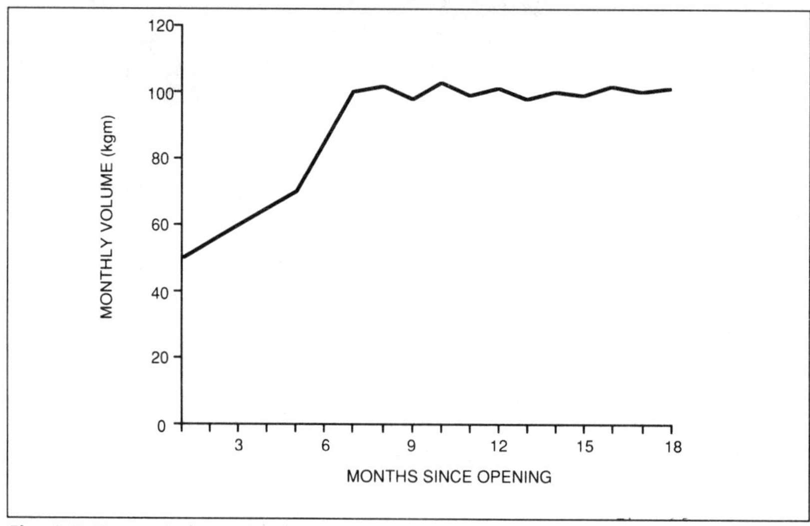

Fig. 6–5 New-to-Industry Station Volume (by Month).

There is also strong loyalty to grade. People who buy regular almost always buy it, and the same holds true for intermediate and premium. Among the grade loyal consumers, premium consumers tend to be of most interest because of the higher margins for retailers. Premium consumers tend to believe in the enhanced performance of the grade, forming a "performance loyal" segment. For these consumers, premium gasoline provides discernible benefits that justify the additional cost. Discernible benefits may come from the "increased pep" or decreased knocking of burning premium gasoline. Other benefits tend to be focused on the car itself, with premium consumers tending to believe that the better grade of gasoline will be better for their cars (see Fig. 6–6).

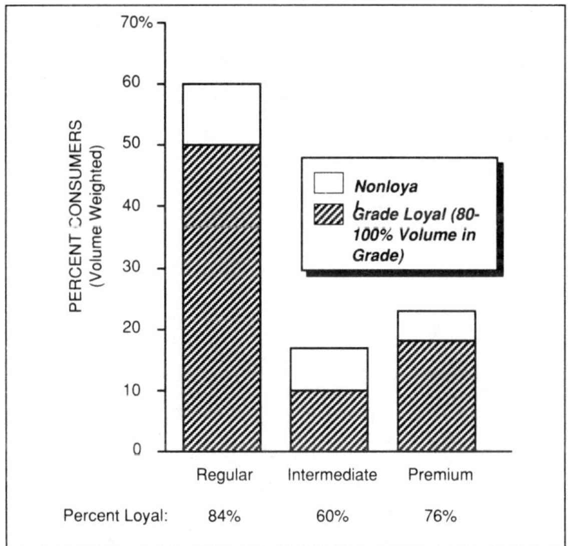

Fig. 6–6 Consumer Grade Loyalty by Grade.

Premium consumers also tend to be more brand loyal than other consumers, a critical consideration for marketers, as shown in Figure 6–7.

Given the potential attractiveness of premium customers, a logical question is, "how large is the performance-loyal consumer segment?" The answer depends on company and even on geography. Some parts of the country, notably the northeast and southern California, tend to sell more premium gasoline than elsewhere. Also, some companies have demonstrated an ability to attract premium customers, even if they tend to market in "nonpremium" areas. Figure 6–8 shows this ability to sell a rich grade mix greatly supports the profitability of these companies.

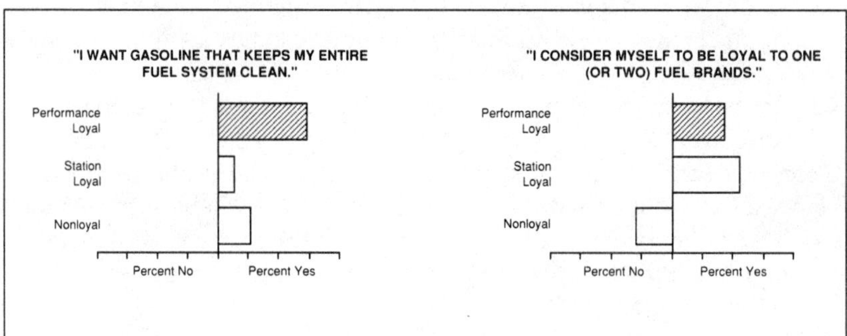

Fig. 6–7 Customer Loyalty.

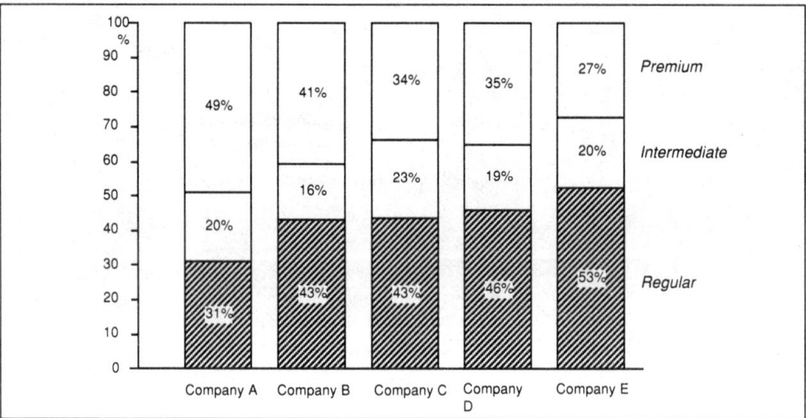

Fig. 6–8 Average Grade Mix (by Volume).

Having discussed the competition for premium customers in some detail, let us point out that other successful marketing strategies are available.

In general, the major integrated companies have the highest costs and fight for the premium customers. Companies can successfully pursue regular grade customers but they must have the cost structure to match. Some smaller marketers (the so-called "independents") have combined high volume, regular focused marketing strategies with lean costs and had spectacular success. The critical lesson for marketers is to match their offering to their customers, because any excess costs in gasoline retailing almost always lead to lower returns.

GASOLINE RETAILING INDUSTRY STRUCTURE

Traditionally, the different types of companies in gasoline retailing could be classified by asset ownership, as Figure 6–9 shows.

Integrated companies owned terminals, trucks, and stations. This tightly integrated ownership allowed companies a high degree of control over what happened to their gasoline.

In contrast, distributors or jobbers owned stations and trucks, but not terminals. Distributors contracted for supply from integrated suppliers. These players typically serviced regions—such as rural areas—where integrated companies were at a disadvantage. An integrated company in a rural area might only have a handful of stations, whereas a distributor could service multiple brands, as well as bulk buyers of unbranded gasoline. Finally, some players owned the service station only. These companies contracted for delivery either from a distributor or from an integrated company.

The unbranded rack is the starting point for gasoline retailing. Transactions at this point are between oil companies and distributor/jobbers or independent unbranded convenience stores. Distributor/jobbers usually resell the product to bulk commercial accounts for which brand is not important (e.g., car rental offices). Unbranded convenience stores leverage unbranded product into a low-priced retail gasoline offering combined with a convenience store.

The point of sale from integrated oil companies to their branded distributors/jobbers is the branded rack. Distributors "brand" their stations with signs and other improvements to market their product to the public.

	TERMINALS ➤	TRUCKS ➤	STATIONS	COMMENTS
Oil Companies	✓	✓	✓	Most stations operated by lessee dealers
Distributors		✓	✓	Focus on rural areas
Contract Dealers			✓	Typically sole proprietorships with 1 station

Fig. 6–9 Asset Ownership.

Typically, at least part of this identification cost is borne by the supplying company. In return, distributors must buy product from the supplying company (since their stations carry a branded sign, the branding company wants to bind the distributor to sell only his fuel). The contracts between distributors and supplying companies typically cover a period of years.

Distributors own and operate service stations and trucks. They obtain product at the rack and deliver it to service stations and bulk accounts. They tend to retail gasoline in rural areas, where they can achieve more efficiency than oil companies by carrying multiple brands and serving bulk accounts as well as service stations. Many also own their own secondary storage facilities.

In most urban and suburban areas, the dominant form of gasoline retailing is the oil company lessee dealer. Here, the oil company's role is greatly expanded from that of distributor-operated stations. The oil company owns trucks or contracts for the delivery of product to the station. In addition, the oil company owns the facility and leases it to a dealer. The oil company sells the product to the dealer at a wholesale price known as Dealer Tank Wagon (DTW). DTW is usually 7¢–20¢/gal above the branded rack price.

Dealers typically have two agreements with a major company. The first is a lease that covers the dealer's use of the station facility, stipulating that a dealer will pay a rental fee for the term of the lease. There are many methods for calculating a dealer's rent. For example, rental fees may be based on the market value of the station, or on the performance of the businesses on the site. The other agreement between dealers and suppliers is a product franchise agreement covering the sale of gasoline by the company to the dealer. The important aspect of a typical franchise agreement stipulates that the dealer will buy fuel exclusively from the supplying company that owns the station. Franchise agreements also usually specify business practices (e.g., cleanliness, customer service) that are consistent with support of the supplying company's brand value.

There is an nascent trend toward augmenting the two agreements between dealers and suppliers with an additional "business methods" agreement. The existing typical product franchise agreement focuses on the product (gasoline) rather than the method or practices used to merchandise and sell it (retail operations). This is a legal distinction that essentially judges the value that the supplier brings to the relationship to be tied mostly to the product rather than to retail practices. In contrast, most fast-food franchises are "format franchises" (i.e., the value in a hamburger is not the meat itself but how it is prepared, packaged, and sold). Some oil companies have moved or are moving toward similar format franchises, notably Arco on the west coast with its AM/PM stations. In this case, there is a format franchise because the AM/PM store configuration, product offering, etc., are sufficiently unique. More of this type of agreement is likely, as oil companies try to get more value out of their station investments.

The preceding discussion focuses on the oil companies, but dealers also have institutional power in the gasoline retailing industry. Dealers' independence and, to some extent, security are protected by the Petroleum Marketing Practices Act (PMPA). PMPA protects dealers by constraining the owning company's ability to act, e.g., a company cannot sell a dealer-operated station until the dealer has had an opportunity to make or match an offer for the station. In addition to PMPA, dealers have formed cooperative organizations to share information and to influence oil company policy. Oil companies are cognizant of the potential constraints posed by their dealers in forming their marketing plans.

Some dealers own the service station themselves and purchase gasoline from oil companies at a delivered price. These dealers are contract dealers. These dealers are typically entrepreneurs who have purchased a service station, often from a major oil company. Contract dealers tend to operate a major related business in addition to gasoline retailing:

- They have large service garages and rely mostly on the auto repair business, but also want to sell gasoline.
- They have convenience stores that wish to offer branded gasoline.

The contractual arrangements for a contract dealer are a blend of those for a distributor and a regular dealer. Like a distributor, a contract dealer has identification costs which are usually funded, at least in part, by the supplying company. Like a regular dealer, contract dealers usually sign a product franchise agreement with the supplying company (although they don't sign a lease since they own the station). Similarly, the price at which contract dealers buy product is a blend of the prices paid by distributors and regular dealers. It is almost always between the branded rack and DTW. Typically contract dealers negotiate a discount off DTW and index their purchase price to that benchmark.

Contract dealers normally work with integrated oil companies in urban areas. In rural areas that are not directly served by integrated companies, branded distributors/jobbers serve contract dealers. The arrangements are similar in nature to those between a contract dealer and integrated company.

The remaining form of service station operation is the oil company's direct sales channel. With these company-operated stations, oil companies are integrated from the refinery all the way to the customer. There are three reasons for oil company's to operate stations directly:

- To retain control of the retail price of the product
- To capture full value from new, high volume sites if prevailing dealer rents are too low
- To stay "close to the customer" and experiment with alternative retailing approaches

In summary, while the integrated oil companies still have a highly visible role across the spectrum of gasoline retailing activities, in some areas a number of smaller companies also operate, sometimes very profitably. There are many models for service station operation, including companies owning no assets but contracting for the operation of entire chains of stations. In this sense the industry, while very mature, still remains highly dynamic. Figure 6–10 shows today's structure of the industry.

There is a substantial variation in the quality of stations operated by integrated companies, dealers, and distributors. In general, major oil companies buy the most expensive locations, build the most expensive stations, and operate them at high volume. These stations are usually a fairly small percentage of a major company's distribution. Dealer stations tend to sell less volume, and to be older stations. Distributor stations are often located in rural areas and tend to pump the least gasoline per station. The profile in Figure 6–11 is typical for a major company: two-thirds of the outlets are

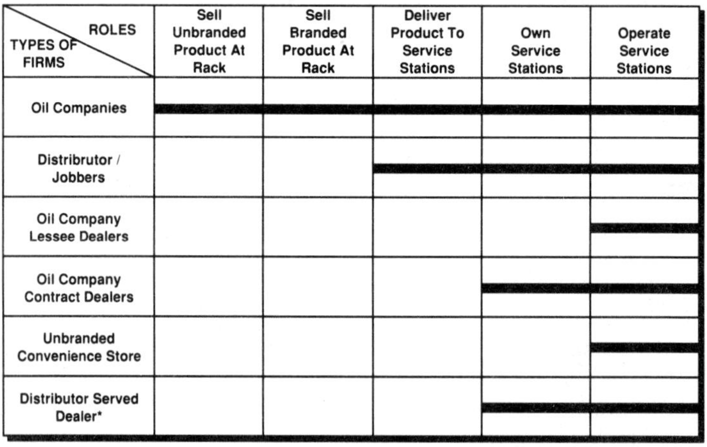

ROLES / TYPES OF FIRMS	Sell Unbranded Product At Rack	Sell Branded Product At Rack	Deliver Product To Service Stations	Own Service Stations	Operate Service Stations
Oil Companies					
Distribrutor / Jobbers					
Oil Company Lessee Dealers					
Oil Company Contract Dealers					
Unbranded Convenience Store					
Distributor Served Dealer*					

Fig. 6–10 Gasoline Industry Structure.

	PERCENT OUTLETS	PERCENT VOLUME	AVERAGE MONTHLY VOUME/OUTLET (Gallons)
Salary Operated	6%	15%	140,000
Dealer/Lessee Operated	28%	42%	90,000
Branded Jobber	66%	43%	40,000
Total Branded	100%	100%	60,000

Fig. 6–11 Major Brand Channel Mix. Source: NPN.

branded distributors (perhaps 4,000–6,000 stations), but they sell less than half the company's total branded sales volume.

TRENDS IN INDUSTRY STRUCTURE

ver time, there has been an unbundling of the gasoline business and a proliferation of different types of players. The gasoline retailing industry structure today is a comprised of five primary functions performed by at least eight types of firms (see Fig. 6–12).

FUNCTIONS:	Sell unbranded product at the rack Sell branded product at the rack Deliver product to service station Own service station
PLAYERS:	Oil companies (major and independent) Distributor/jobbers Oil company lessee dealer Oil company contract dealer Unbranded convenience stores Distributor/jobber-served dealer Terminal operation specialists Station operation specialists

Fig. 6–12 Retail Gasoline Industry: Primary Functions and Players.

The trend in industry structure has been toward smaller, focused companies, rather than the national integrated chains. Starting at the terminal, specialist operators, notably GATX, have played a consolidator role of owning and operating gasoline terminals, supplying multiple brands from a single facility. At the other end of the chain, multistation contract dealers have become common. These companies have been able to negotiate competitive delivery contracts from supplying companies while they focus exclusively on retail operations.

Other companies have been successful focusing geographically. Some distributor chains now include several hundred stations and wield considerable power with supplying companies. As a portion of total service stations in operation, distributors have grown rapidly over the past 20 years.

Some integrated companies also have been successful focusing geographically. For example, on the West coast, Arco has combined tight vertical integration from its low-cost Alaskan crude with its AM/PM convenience store format. In the Midwest, Ashland Oil's SuperAmerica chain combines a novel convenience store format with low-priced gasoline. Most of the SuperAmerica chain is company operated, both to maintain quality control and to recapture the full value of the franchise.

Similarly, several major oil companies have focused their marketing efforts geographically by trading stations with other companies to improve their position within a market. For example, Mobil is now mainly an player on the East and West Coasts, while Amoco is the dominant major in the Midwest.

The move toward "small is beautiful" in the gasoline retailing industry is likely to continue. There will be increasing pressure on traditional integrated structures by nimble and efficient specialists. In time, some of these specialists may become a dominant force in the industry.

STATION FORMATS

oncepts and practices in the physical configuration, or format, of gasoline fueling facilities have changed substantially over the past 20 years. This change has opened opportunities for new companies since it can be difficult and prohibitively expensive to convert an existing chain from one format to another. The search for the "format of the future" goes on—most likely it will include more businesses in addition to basic gasoline retailing.

BACKGROUND

The traditional format for a gasoline fueling facility has been the service station. Service stations supplied a full range of services for automobiles, including fueling, maintenance, and repair. Consequently, the typical station format included one or two islands with pumps for fueling, with a like number of service bays for maintenance and repair work. Station employees performed all levels of service at the station, from fueling the car to major mechanical work. In fact, the term "service station" is still used generically to describe any facility that sells gasoline.

Three changes in the industry combined to make the traditional two-bay format relatively disadvantaged:

- The move to self-serve islands
- The escalating sophistication and proliferation of automobiles
- The increasing desirability of locating gas stations on high traffic corner locations

The move to self-serve islands was an interesting innovation that backfired in many ways for the oil companies. The practice originally was intended to provide a lower cost option for customers during the period of escalating gas prices in the 1970s. But in fact it broke the traditional spectrum of service provided (see Fig. 6–13). Station attendants used to fuel the car, check wipers and fluid levels, and recommend needed service. Self-service erased that interaction between the customer and station employee as people learned to fuel cars themselves. Without the regular links between fueling, monitoring, maintenance, and repair, service stations became "gas stations."

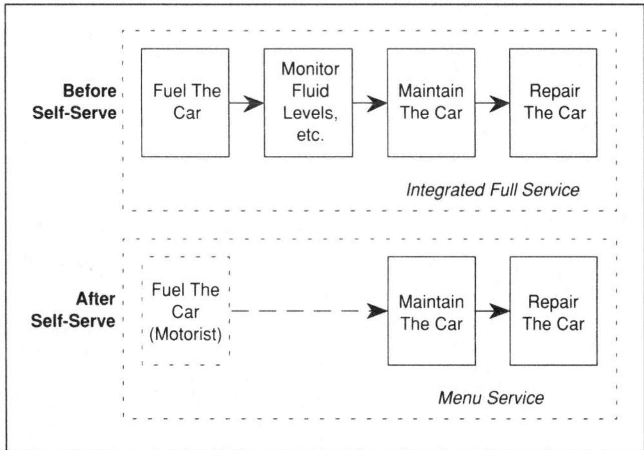

Fig. 6–13 Self-Service: Before and After.

The second major change in the auto service business was the proliferation and increasing sophistication of cars. In the early 1970s, the vast majority of cars sold in the United States were made by the Big Three auto makers, and each company had a limited number of models. This made it relatively easy for a two-bay station to stock in inventory the most needed parts and to train its employees on maintenance and repair procedures.

With the increasing preference for foreign cars and the increase in models offered by the Big Three, the number of potential parts required in inventory multiplied dramatically. Furthermore, cars became more mechanically sophisticated (for example, in terms of electronics). The need for greater depth of parts and service made size more important. Successful auto repair facilities tend to be large so they can cater to multiple customer needs efficiently.

In addition, complexity of cars led auto makers to claim that they best knew how to maintain and repair the cars they made, often offering guarantees as proof. Specialist repairers such as Midas Muffler could make similar claims within their own area of expertise. Figure 6–14 outlines the competitive factors at work in auto service.

The third major change affecting the bay business had more to do with commercial real estate than with the oil industry. Building a service station is expensive, and a station located on a busy street or corner does a lot more business than one on a side street. Increasing regulatory requirements, and the demands of developers and zoning boards, increased the cost of building a station, which made higher sales volume more important in recouping the investment. As a result, companies increasingly preferred to locate their stations on high traffic corners to maximize gasoline sales. This put them in direct competition for location with the growing convenience store and fast-food industries, which also craved high traffic locations to maximize sales. However, the convenience in fueling at a high-traffic location did not fully transfer to auto repair. People wanted convenient locations to make quick, relatively small purchases that would not interfere with their schedules. Auto repair was by definition a major interference. Once again, the traditional service bay was put at a disadvantage.

By far the most popular format alternative to the traditional bay station has been a combined fueling facility and convenience store (see Fig. 6–15). Both of these businesses benefit directly from increased traffic flow. Moreover, these are complementary businesses in that most purchases are relatively small and there is little customer service involved. As a result, labor can be shared across the businesses, allowing a single attendant to run an entire facility. Oil company convenience stores (c-stores) typically have substantial fueling facilities to handle peak hour traffic. All major oil com-

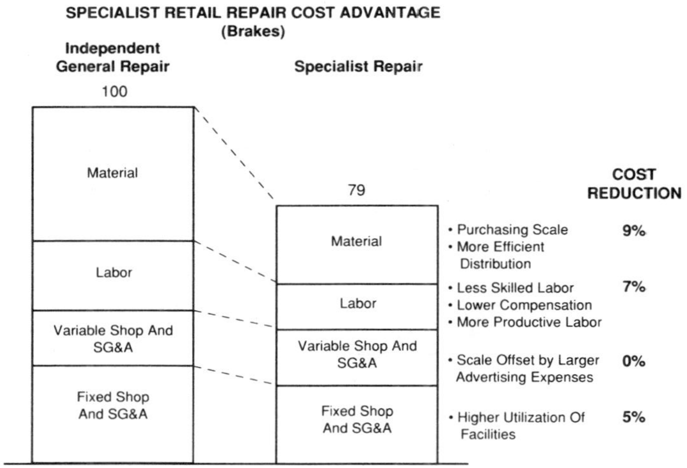

SPECIALIST RETAIL REPAIR COST ADVANTAGE
(Brakes)

Independent General Repair
100

Material
Labor
Variable Shop And SG&A
Fixed Shop And SG&A

Specialist Repair
79

		COST REDUCTION
Material	• Purchasing Scale • More Efficient Distribution	**9%**
Labor	• Less Skilled Labor • Lower Compensation • More Productive Labor	**7%**
Variable Shop And SG&A	• Scale Offset by Larger Advertising Expenses	**0%**
Fixed Shop And SG&A	• Higher Utilization Of Facilities	**5%**

SOURCES OF COMPETITIVE ADVANTAGE

CUSTOMER SERVICE	OPERATIONS
• Brand Image Supported By Large Scale Advertising - National Scope - Regional Dominance • Large Facilities Wth Standardized Services • Local Ubiquity In Major Metropolitan Markets • Available, Low Cost, Quality Parts	• Low Cost Parts Procurement • Effective And Low Cost Parts Distribution • Low Skilled Labor • Sophisticated Information Systems - To Manage Parts Replenishment - To Control/Monitor Shop Performance - To Execute Marketing Programs

Fig. 6–14 Competitive Factors in Automotive Service. Source: BA&H estimates.

FORMAT	NONGASOLINE GROSS MARGINS (% OF TOTAL)	% OUTLETS	% GASOLINE
Bay Store	36%	41%	34%
Convenience Store	26%	35%	32%
Motor Fuel Only (MFO)	10%	14%	19%
Interstate Stores*	N/A	10%	15%
		100%	100%

Fig. 6–15 Gasoline Station Formats.

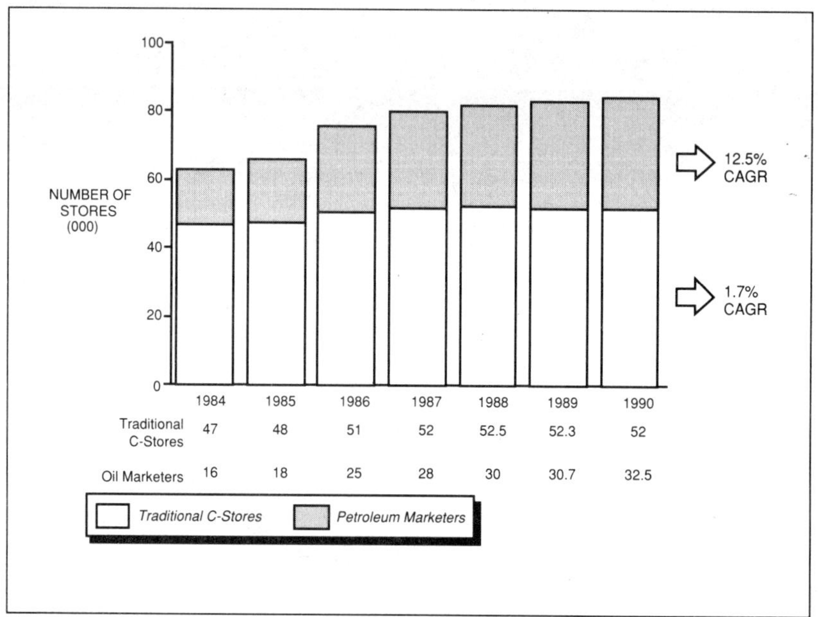

Fig. 6–16 Growth of Petroleum Marketers in C-Store Business.

panies and most large independents have developed a convenience store concept, as shown in Figure 6–16.

Even as the oil companies have tried to become convenience store operators, the convenience store companies have added a gasoline offering. Today, it is uncommon to visit a c-store that does not have a fueling facility, usually offering its own brand or a lesser known brand. By mainly offering unbranded gasoline or gasohol, c-store companies have been able to price their fuel below the oil companies. In fact, it is arguable which group (the oil companies or the c-store companies) have made the better transition to the other's format. Clearly, the oil companies are trying to break their gasoline mentality and become more customer oriented, like their c-store industry competitors (see Fig. 6–17).

The independence of the gasoline and c-store businesses is counter–intuitive. After all, combined c-store/gas stations typically share a brand sign, and location, and the same personnel. However, Figure 6–18 shows that correlation of c-store sales and gasoline sales is poor.

While there appears to be some crossover of gasoline customers to get convenience store items (for example, getting a soft drink or coffee at time of fill-up), customers who travel to convenience stores very seldom buy gasoline. This lack of correlation reveals a basic foundation of gasoline retailing today: Successful gasoline retailers are also good retailers of other products. The need for multiple skills is driven by the need to get the maximum return from high-traffic real estate. Land values reflect a lot of potential sales volume from high-traffic lots. Companies need sophisticated marketing capabilities and the right portfolio of businesses on a site to make a profit (see Fig. 6–19).

Not all new formats have had the same success as the c-store. Oil companies have experimented with stations that sell gasoline only, in multiisland configurations commonly called "pumpers." As a group, these locations have not yielded sufficient income. The reason appears to be that while a given location may not have enough gasoline traffic, development can be justified if multiple business can be sited on the same spot. In other words, the "highest and best use" of many high traffic corners that drives their value is a multibusiness facility which may include fueling.

The cost of land and need to rethink the businesses on site has led some oil companies to correctly shift their focus from gasoline to real estate. The economics are fairly compelling: attractive sites average around $750,000 to buy, and development costs range from $600,000 to $1,000,000, depending upon the type of station built. Using a typical pretax return of 18% as a hurdle rate, new stations need to clear $250,000–$300,000 just to cover their annual carrying cost of capital.

	EXXON	AMOCO	CHEVRON	MOBIL	SHELL	TEXACO	ARCO
"Implied" Objective	←Provide Incremental Margin to Core Gasoline Retailing Business→						Separate Core Business From Gasoline
Store Name	Exxon Shop/ Tiger Mart	Food Shop	Food Mart	Mobil Mart	Food Mart	Food Mart, Star Mart	AM/PM
Positioning	←C-Store not Positioned as Destination Apart From Gasoline Retailing→						Destination C-Store
Primary Store Design	1,100-1,400 ft? Under Canopy	900 ft? Under Canopy	1,000 ft? Under Canopy	1,000 ft? Under Canopy	800-1,400 ft? Under Canopy	1,100-1,400 ft? Under Canopy	Behind Canopy, 2,000 ft? Average
Product Mix	←Cigarettes, Beer/Wine, Soda Account for Over 50% Sales, Trying to Increase Food Sales→						More Weighted Toward Food Service
Pricing	At or Below Average	N/A	N/A	N/A	N/A	N/A	Low on Some Items, High on Others
Other	Building Larger Stores	Rolling Out Split Second Stores		Testing Co-Branded Food Products	Testing New "ETD" Concept	Testing Co-Branded Food Products	Expanding AM/PM to International Markets

Fig. 6–17 Major Companies Retail Format Summary.

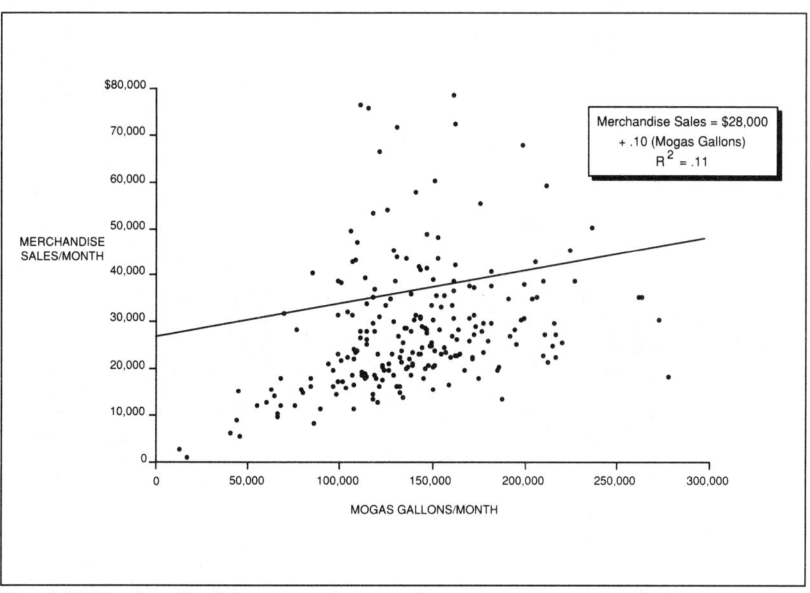

Fig. 6–18 Merchandise Sales verses Mogas Volume. Source: BA&H analysis.

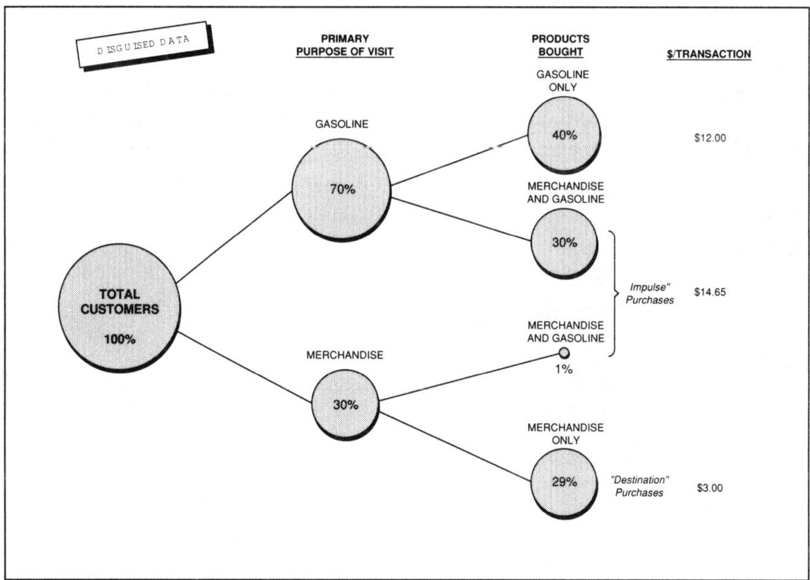

Fig. 6–19 Buying Patterns of Station Customers.

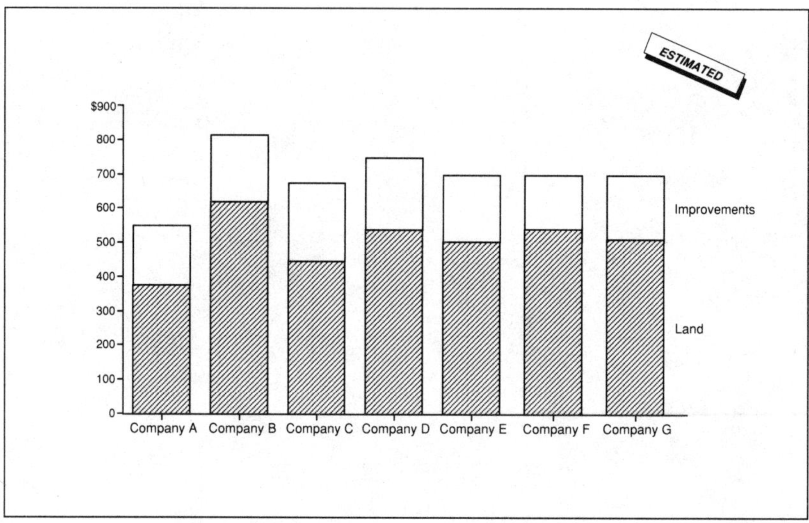

Fig. 6–20 Station Values (1991).

Clearly, as Figure 6–20 shows, companies need to maximize the cash flow from each site to justify high development costs.

TRENDS

The current trend in station formats is to locate as many convenience-related businesses on site as possible. These often are branded businesses, resulting in multiple brand identification signs at a single location. Current offerings beyond a fueling facility include (but are not limited to!):

- Convenience stores
- Car wash
- Fast food (hamburgers, pizza, yogurt, etc.)
- Banking (ATM)
- Postal services
- Video rental

This trend in multiple business offerings seems likely to continue as long as all of these types of companies are vying for the same real estate.

An alternative format outside high traffic locations is the automated fueling facility. This is the "pure" gas station, with no attendant and credit

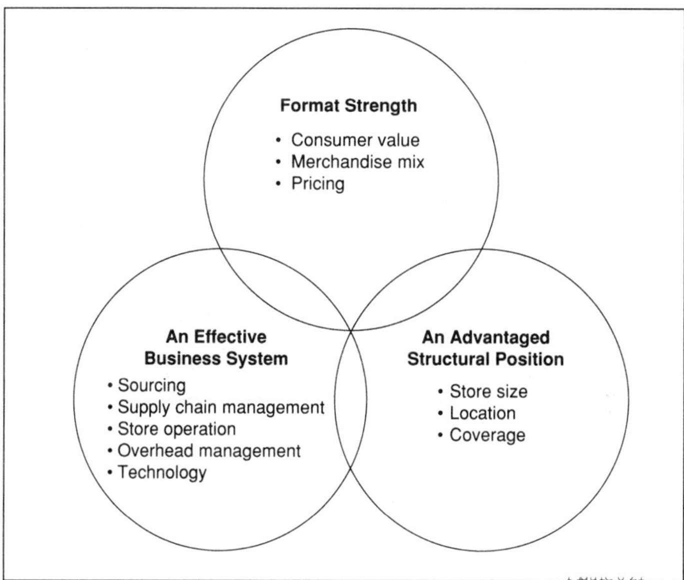

Fig. 6–21 An Integrated Approach to Format Development.

or debit card-only sales. Until now, this format has been used mostly with commercial accounts that are interested mainly in low costs and control, but several companies are experimenting with broader offerings of this format as well.

History has shown frequent changes in retailing formats, and gasoline retailing is unlikely to prove an exception. As advantaged formats evolve, companies with existing chains will be faced with the painful choice of shifting formats to stay current or to make do with their existing offering. Given the high cost of development, it would seem that in the future, companies would place a high premium on flexible formats that could be reconfigured at a reasonable cost (see Fig. 6–21).

RETAIL OPERATIONS

While gasoline retailing outlets are owned by different types of firms, the underlying retail economics are somewhat similar. Service station economics are, in essence, the sum of all businesses on the site. Each business has its own margin, yet businesses often share the costs of operation (see Fig. 6–22).

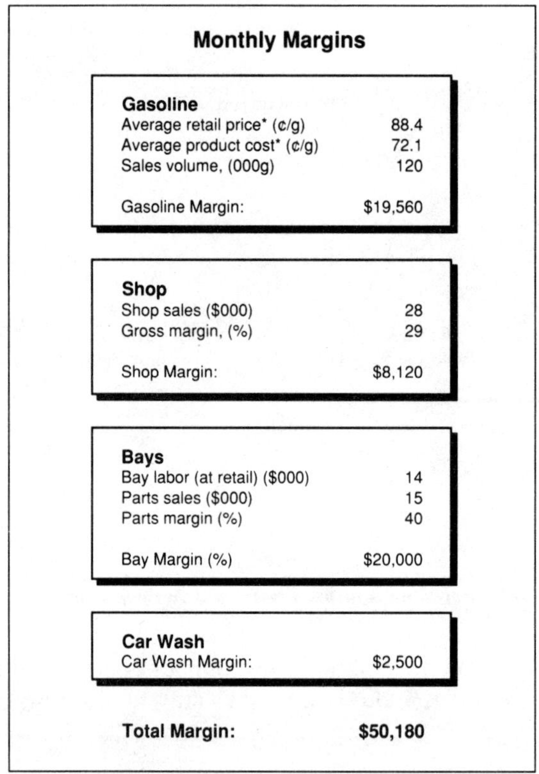

Fig. 6–22 Typical Retail P&L for a Company Operated Service Station.

Gasoline margin is the product of gasoline volume and unit margin. Unit margin in turn is the difference between retail price and product cost. Unit margins vary substantially by grade and by level of service (full service is much more profitable than self-service). A company's grade mix is a key driver of its gasoline profitability.

Factors which determine gasoline volume are the size, location, and appearance of the facility; the number of competitors in the immediate area; and the hours of operation. Each of these factors can substantially affect gasoline sales volume. Figures 6–23 and 6–24 outline some of these impacts.

Bay margins represent the markup on parts installed in vehicles and the retail value of the labor services provided. Typical margins range from $7,000–$20,000 per bay per month.

Shop margins are the gross profits from sales of snacks, beverages, tobacco, and convenience items. Even stations without shop facilities will typically sell some merchandise. A typical oil company shop facility will sell $25,000–$40,000/month with gross margins of 25%–35%. Convenience stores, offering unbranded gasoline, often have much larger facilities with more extensive product offerings. For them, monthly sales of $60,000–$80,000/month are common.

Car Wash margins are the revenues from providing car washes. These revenues are a function of the sophistication of the equipment. Service offerings range from short track soap-and-spray machines to tunnels equipped with polish, wax, and corrosion inhibitors.

Retailing expenses are dominated by labor. The labor budget consists of the dealer's income and/or manager's salary, cashier wages, and mechanic's wages (for bay stations).

Staffing a gas/shop combination is relatively simple. Stations require one employee at all times. Additional staff is needed in very large stores or during peak periods. Managing a station is generally not a full-time job, so managers serve as cashiers.

Bay labor management is more complex. Demand is less predictable. Also, mechanics usually earn more than cashiers and have different skills, so the operator's task is to match work with the lowest wage mechanic who can do the job properly.

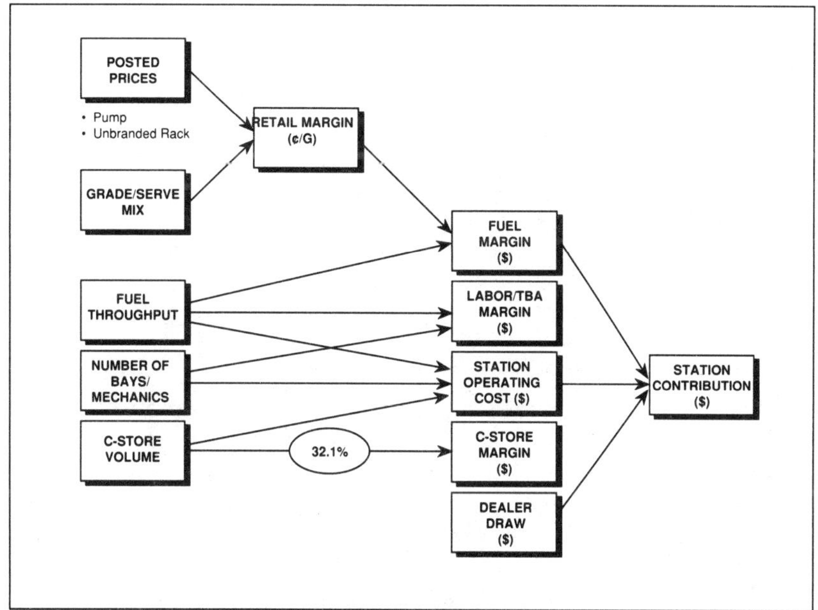

Fig. 6–23 Estimating Station—Level Contribution.

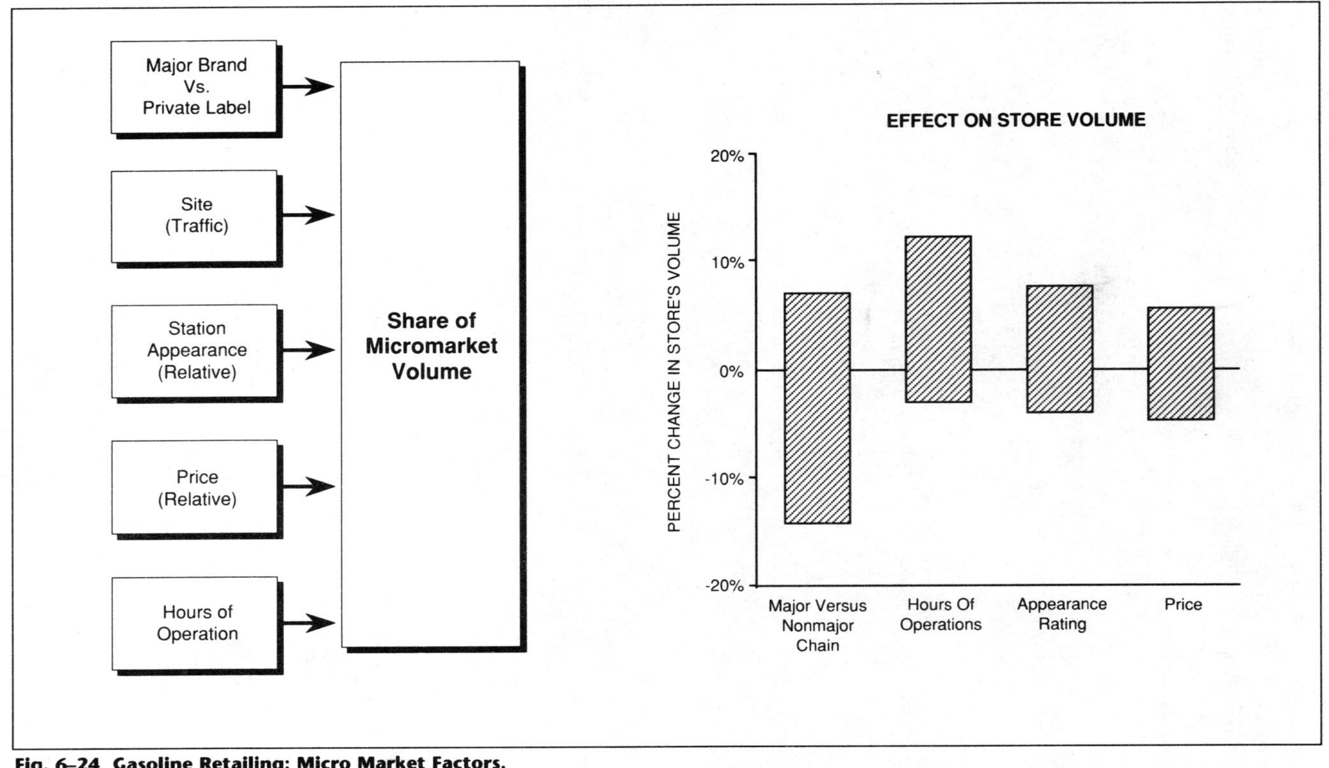

Fig. 6–24 Gasoline Retailing: Micro Market Factors.

Landlord expenses are paid by the facility's owner and include maintenance, property taxes, depreciation and lease expense, if the property underlying the station is leased rather than owned. If an oil company leases the facility to a dealer, it generally views rent income from dealers as an offset to these landlord expenses.

Given all this, it would appear that gasoline retailing is a very profitable business, or at least it can be. On the other hand, the factors discussed in this chapter highlight the dimensions of performance that can vary from company to company and from station to station. On average, gasoline retailing is a marginal business, where independent dealers work long hours to make a relatively modest living. However, some companies have achieved substantial returns through a combination of careful management and savvy marketing.

Figure 6–25 shows a estimate of operating return on capital for the major oil companies (Amoco, Chevron, Exxon, Mobil, Shell, and Texaco). As the graph shows, gasoline retailing has been a marginal business for most of these companies, but as always, there is an exception.

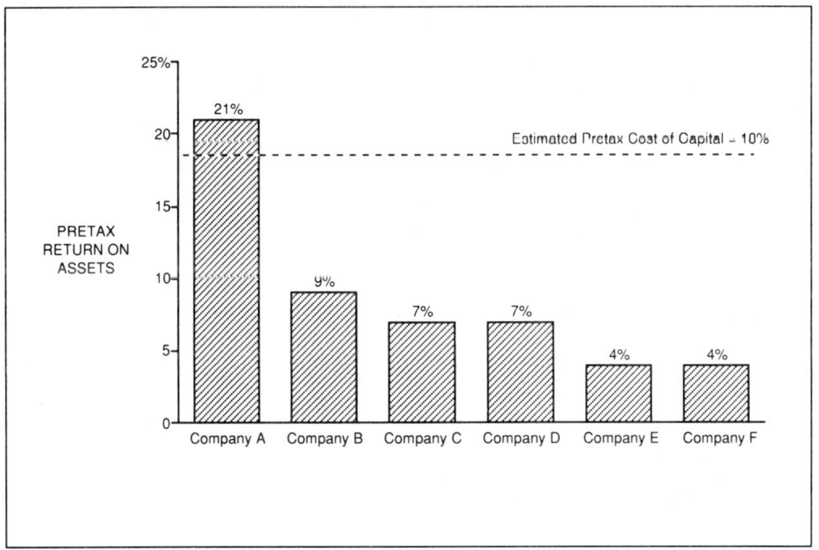

Fig. 6–25 Operating Return to Capital (1988-1992). Sources: Whitney Leigh; MPSI; OPIS; New Image; BA&H analysis.

SECTION III
BUILDING STRATEGIC CAPABILITIES

CHAPTER 7

BUSINESS PROCESS DESIGN

As management consultants what we see most frequently is not a fundamental failure of strategy selection, but failure of execution. In this section we show how to build a company's capabilities to best implement the selected strategies. The discussion starts with today's hottest topic—business process redesign—and then moves on to discuss organizational design, planning, materials management (sourcing and maintenance), quality management, human resources, and the role of the extended enterprise.

It has been called design, redesign, engineering, reengineering, and in one flight of fancy, "the only hope for breaking away from the ineffective, antiquated ways of conducting business that will otherwise inevitably destroy."[1] Destroy what? With that kind of fanfare, it sounds as if it may be too late for those of you reading this chapter if you are not already deep in the midst of a massive reengineering program.

On the other hand, there is a saner way to look at this phenomenon. Rather than being the do-or-die revolution that will separate the quick from the dead, we will argue that business process redesign (BPR) is a natural evolution in management thinking and as a result, can be accomplished without the tremendous upheaval that "revolution" implies. Further, oil and gas companies have already been using BPR (in some cases without the names and hoopla) in order to reduce costs in response to the price shocks of the 1980s. Finally, we will show that BPR's highest calling may be in helping oil and gas companies design their organizations for the 1990s and beyond.

FUNCTIONAL ORGANIZATIONS AND THE BIRTH OF BPR

In this section we will examine the earlier claim that BPR is a natural evolution in management thinking. To do this, we will briefly describe the development of the functional organization, the dominant form of business in the latter half of this century. We will also show that for oil companies at least, the functional organization became a barrier to the kind of quick restructuring that was necessary to survive in the mid-1980s. As part of that discussion, we will highlight some of the other issues raised by the functional organization—most notably, its inability to move quickly or to serve customers adequately. Finally, we will present BPR as the natural solution to these problems.

An easy way of describing organizational development over the last 50 years is to set up the distinction between the vertical (functional) organization and the horizontal (process/product) organization. Figure 7–1 shows the difference.

The functional organization relies on the specialization of its participants: putting together the engineers, the explorationists, the accountants, and others.

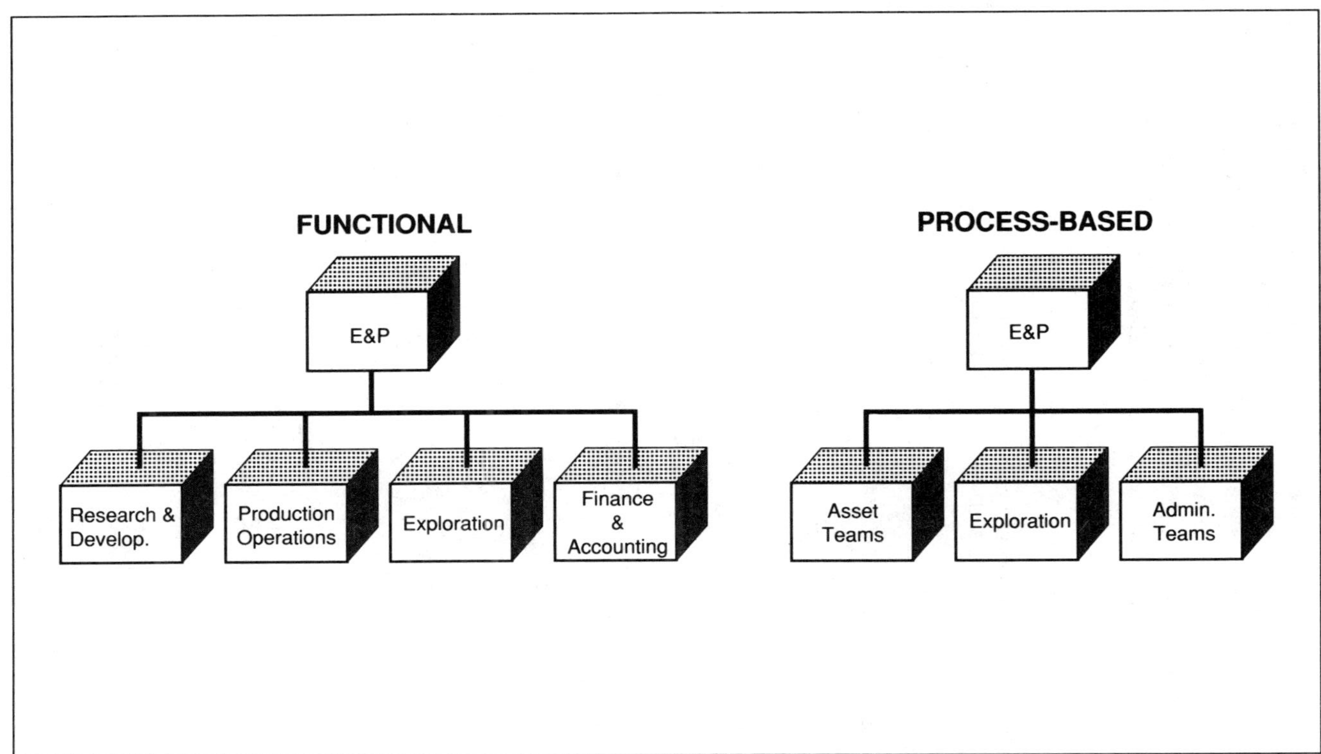

Fig. 7–1 Functional vs. Process-Based Organizations.

The horizontal or process-based organization in an oil company can take on several forms. Upstream, or in refineries, it may be the asset team, a multifunctional group that concentrates on a producing field or a refinery unit. Administratively, there may be a pure process organization, for example, putting land, land administration, division of interest, and accounting people together to concentrate on all land-related processes.

Each organization has its advantages, as shown in Figure 7–2.

• Maximum development and utilization of specialized skills	• Better coordination and integration of work
• Cost effective divisions of labor	• Quicker response times
• Economies of scale in plant and equipment	• Simpler cost controls
• Convenient development of centralized coordination and control	• Higher levels of creativity
• Effective hiring of, and clear career paths for, specialized experts	• Greater job fulfillment

Fig. 7–2 Left column: Functional Organization Benefits, Right column: Process-Based Organization Benefits.

Further, the distinction between the two is not new. The functional organization dominated in the 1940s. In the 1950s one of the first process organization innovators was Standard Oil. Its chairman, A A. Stambaugh, began flattening organizational hierarchies, pushing decisions down to the lowest level possible, minimizing the use of staff, and maximizing cross-functional teamwork. Of course, at the time it was thought of mostly as decentralization, not BPR.

An interesting side note on the debate was the 1970s innovation: the matrix. Unable (or unwilling) to make a choice between functions and processes, many companies tried a compromise that included both. Once they discovered the endless debates over "hard line/soft line" reporting and the attendant "Who's on First?" mentality that resulted, the matrix fell into the dustbin.

By the 1980s the functional form was dominating most oil companies and most companies in other industries as well. Oil companies pushed hard on specialization. It was no longer just engineering, but reservoir engineering, facilities engineering, etc. It was no longer just accounting but oil revenue accounting, gas revenue accounting, and marketing accounting. The complexity of the oil and gas business led to specialized skills in employees. This specialization was then mirrored in the organization structures of the time.

And then the wake-up call came. The price shocks of the 1980s, well documented elsewhere, caused oil companies to move—and move fast—to stop the bleeding. As they did this, they found that the functional organization was a hindrance to the kind of fast restructuring that was needed for survival.

Why? Faced with a massive loss of revenues, the imperative was cost reduction. And cost reduction, as has been learned, is best accomplished by reducing work. In a functional organization, however, this is nearly impossible because almost no meaningful work is accomplished within a single function. Instead, most work crosses organizational boundaries. Figure 7–3 is a simple illustration of what we mean.

In order to take out any work (and the costs that go with it), careful coordination among all the functions involved is required. And that is difficult and time consuming. In the 1980s, oil and gas companies either did

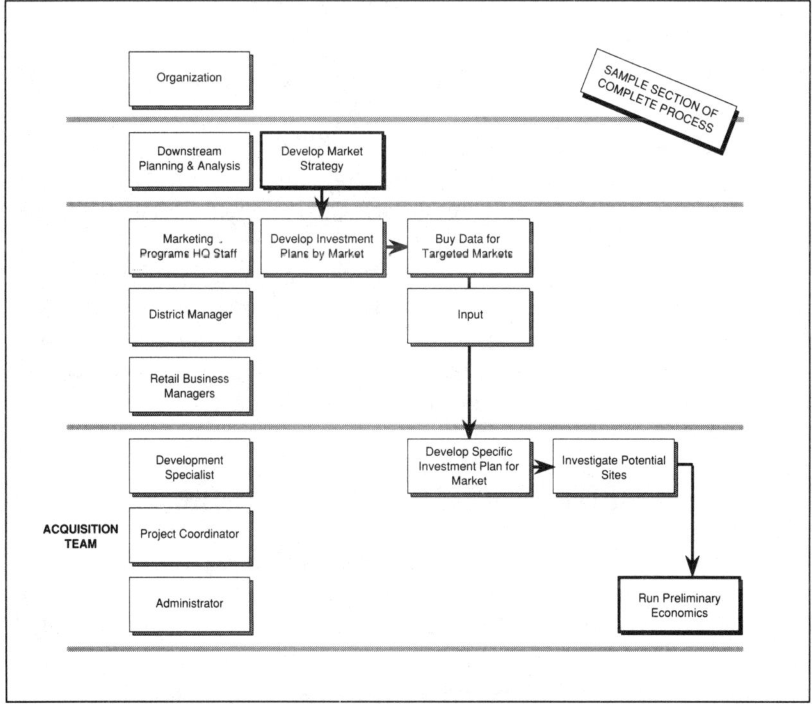

Fig. 7–3 Business Process Re-Engineered Site Selection and Acquisition.

not have the luxury of performing such detailed analysis, or they were unaware of the need for it.

Instead, they relied on across-the-board cost cutting, telling each department to cut, for instance, 20% of their costs in order to bring the bottom line back into the black. And as was discussed in the first section of this book, this kind of cost cutting set the stage for some of the dramatic performance problems the industry is currently experiencing.

Once the across-the-board cuts were made, oil company executives were chagrined to find that the costs came back. Without actually getting rid of any work, companies were forced to recall (sometimes at expensive consultant rates) the employees they had just let go in order to get the work done. All of this occurred, at least in part, because the functional organization chart did not give management a good understanding of what was really going on within the company.

There are other problems with the functional organization. Specifically, it is slow to move (death in today's fast-changing environment), and it is a difficult structure from which to create the "customer-focused" mindset which is once again in vogue. We will consider each in turn.

SPEED IN THE FUNCTIONAL ORGANIZATION

Much has been made lately of the advantages of speed in an organization. Speed means getting new fields on line quicker, getting product to market faster: in short, speeding up the cash flow and reducing expenses by taking steps out of key business processes.

The functional organization is slow. Decisions must pass from function to function, with each function "adding value" by performing its own little piece of the analysis. Focusing on functional interests, we get the phenomenon of "checkers checking checkers." Figure 7–4 shows a typical decision making process in a functional organization.

SERVING THE CUSTOMER

Similarly, the functional organization is difficult for the customer to engage. Rather than having one simple point of contact, the customer must deal with different functions in order to solve problems or communicate. For example, the independent dealers of the typical major oil company face a myriad of contacts to transact normal business:

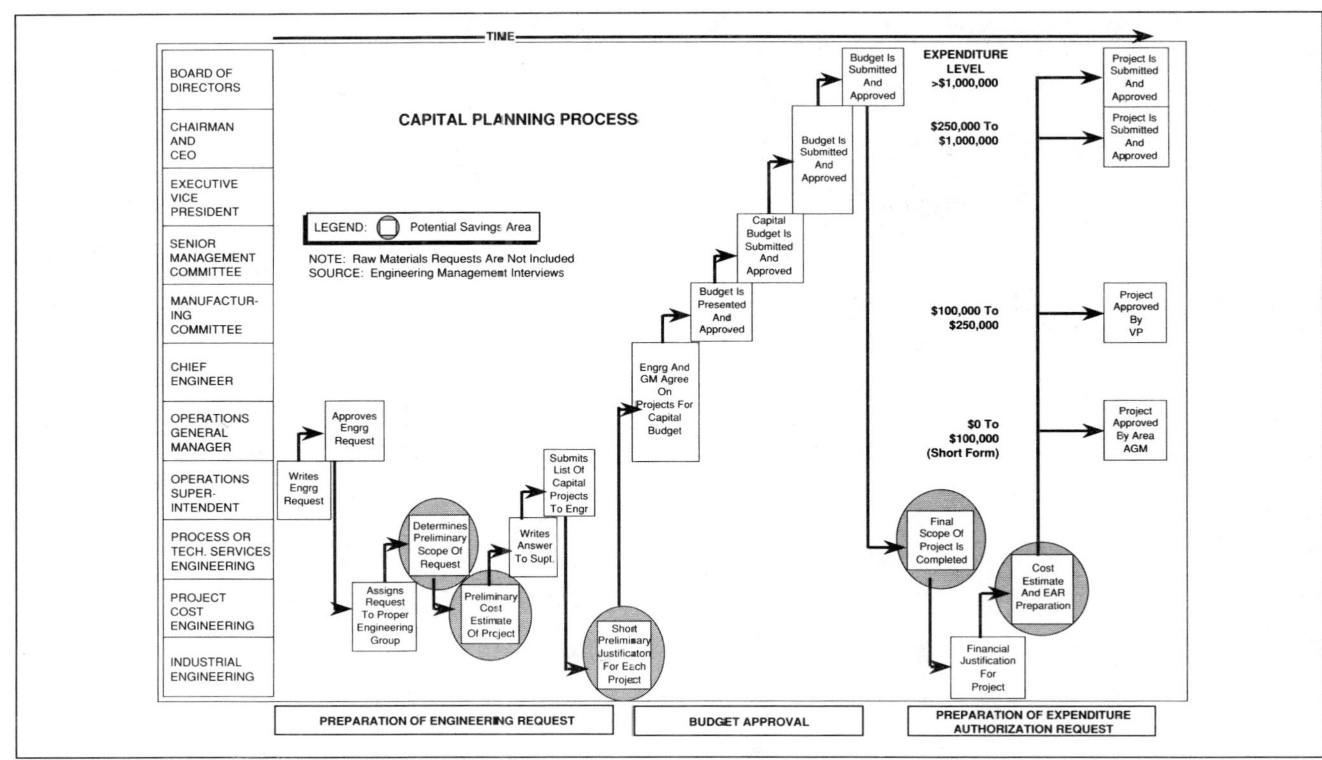

Fig. 7–4 Functional Organization Decision Making.

- One phone number to order gasoline deliveries
- A maintenance group to call for maintenance and repair needs
- An administrative group to call with any billing questions
- A group in charge of special marketing programs to handle promotions and the like
- A credit department to contact in case of any payables issues that delay deliveries
- The terminals to call in case of any delivery scheduling problems
- A field sales force that contacts the dealers and tries to help them navigate the maze of other contacts

It is not surprising that customers get frustrated with the functional organization. Nor should it surprise anyone to find that there are real costs of complexity in this kind of relationship on top of the reduced customer satisfaction.

Difficult to cut costs, slow, unresponsive to customers, the functional organization was clearly a problem for the oil and gas companies in the late 1980s. As a result, many of them started "doing BPR" before the acronym was popular. They did it for the reasons cited above: a desire for lasting cost reductions, a need to move faster, and a requirement to improve customer service.

BPR—WHAT IS IT?

We have described where BPR came from without describing what it is. Perhaps the best way to describe it is by example. (The examples in this chapter come from Booz·Allen and Hamilton client experience with oil and gas companies. In all cases, details have been altered to protect the confidentiality of our clients.)

EXAMPLE 1—SERVICE STATION SITE SELECTION

A major oil company's gasoline retailing subsidiary used BPR to address what it perceived to be a significant cost disadvantage to the rest of the industry. One key process was site selection, picking the right site to build a service station.

The company's strategy had evolved from broad national growth in many markets to more focused growth in a few key markets, but the organization and processes had not fully adapted to the change. As a result, site selection was still handled as if the company were in broad, rapid growth.

A cross-functional BPR team profiled the site selection processes and found major opportunities for improvement. For example, the company had two groups charged with identifying and selecting sites. A centralized headquarters staff made brief field visits, then analyzed site potential in the home office, while local real estate staff focused on understanding local market trends and opportunistically identifying sites. Local real estate staff members also were responsible for ongoing service and maintenance of existing properties. Both groups had to interact with many other functional groups who were involved development and station operations, and they had to have a headquarters committee approved all major decisions.

More than 100 people participated in this process at an administrative cost of over $150,000 per site. Worse, the cumbersome analysis and approval process was taking up to 3 1/2 years to complete. Obviously, by that time all the best sites were gone. The company watched as both traditional competitors and retailers such as the fast-food companies snapped up desirable real estate in growth markets.

Also, as a result of the change in strategy to more targeted expansion, the headquarters group and local real estate staffs were overstaffed for the volume of sites required, and they competed with each other to find sites in the same markets. Because the groups were measured on the number of sites evaluated and not on how well the resulting investments did, the company purchased more sites than necessary.

Processes grew increasingly complex to coordinate between the two groups and the other functions. Benchmarking vs. other companies competing for the same sites revealed that the company's costs and cycle times to obtain sites were frequently twice those of best practices.

To remain competitive, the company needed to dramatically improve performance, setting a target as shown in Figure 7–5.

Local teams were formed in core investment markets and were given the resources and authority to manage the entire process from cradle to grave. They retained the value-added activities performed by headquarters and field staff, but many of the coordinating and analytic requirements were eliminated. This approach was consistent with a broader decentralization of business accountability to regional field-based management. Accountability for investment returns became clearly focused through the teams.

The teams have substantially improved performance, achieving the high targets that were set. The company saved several million dollars per

RETAIL CLIENT EXAMPLE

MEASURE	TODAY	BEST PRACTICE TARGET
TIMING	24-40 Months	12-20 Months
PROCESS COSTS PER SITE	$100-160 Thousand	$30-65 Thousand
ORGANIZATION FTEs (People)	Many (100-150)	Fewer (30-60)
HANDOFFS	80-95	<40
ORGANIZATION UNITS INVOLVED	15-20	5-10
SITE QUALITY	Volume Based	Profit Based
ACCOUNTABILITY	Fragmented	Focused

Fig. 7–5 Site Selection and Acquisition Process Targets.

year while improving effectiveness and speed. The team positions are highly challenging but highly rewarding, because team members have much more authority for getting the job done.

Two interesting notes: First, when it came time to benchmark companies that were finding sites well, the oil and gas company looked not to its peers but to the fast food retailers that were beating them to the punch for the best locations. Also, through BPR they discovered it was possible to outsource many of the real estate transactions to real estate firms, which were better at it.

This example describes all of the elements and benefits of BPR that we have been talking about. The client:

- Took a cross-functional, horizontal look at how work was really getting done.
- Created a target for necessary performance—improvements in speed, cost, and quality
- Used creative ideas from benchmarks and both internal and external sources to lead to a new process
- Implemented a better way to get work done

So how do you really do this?

There are many approaches to BPR in the marketplace. We will briefly discuss the model developed and used at Booz·Allen. Major elements of our

methodology came from our work with oil and gas companies. The price shocks of the 1980s created a real-time development lab for the approach.

Through this client work, we have identified the principles of successful BPR, shown in Figure 7–6.

• Secure top management sponsorship • Guide effort by strategy and the company's capabilities • Develop a communications and change management plan • Set stretch goals as a "stake in the ground" • Focus on business processes regardless of organizational boundaries • Target high priority business processes	• Employ a proven restructuring methodology • Ensure internal ownership of recommendations • Install results oriented performance measurements • Establish an implementation blueprint with clear endpoints/milestones • Assign responsibility for implementation

Fig. 7–6 Booz·Allen's Business Process Redesign Principles.

These principles drive our approach to business process redesign, shown in Figure 7–7. We will discuss each step briefly.

STEP 1 IDENTIFY YOUR CAPABILITIES

What do we mean by capabilities? Capabilities are a unique combination of "know-how" and the systems used to deliver the value of that know-how to customers. Capabilities can include resources, infrastructure, accumulated knowledge, brand equity, business processes all deployed to meet customer demand. In an oil and gas company, they can include oil finding expertise, superior distribution logistics for gasoline, or advantaged production economics, for example.

The first step of BPR is to understand both your current capabilities and those that will be needed to compete successfully in the future. You want to make sure that your business processes are congruent with your strategy.

In the site selection example, the company had decided that location was a key determinant in gasoline retailing success. Further, since there were definite advantages to being "first-on-the-corner," a key capability would be the ability to move quickly to secure the best sites.

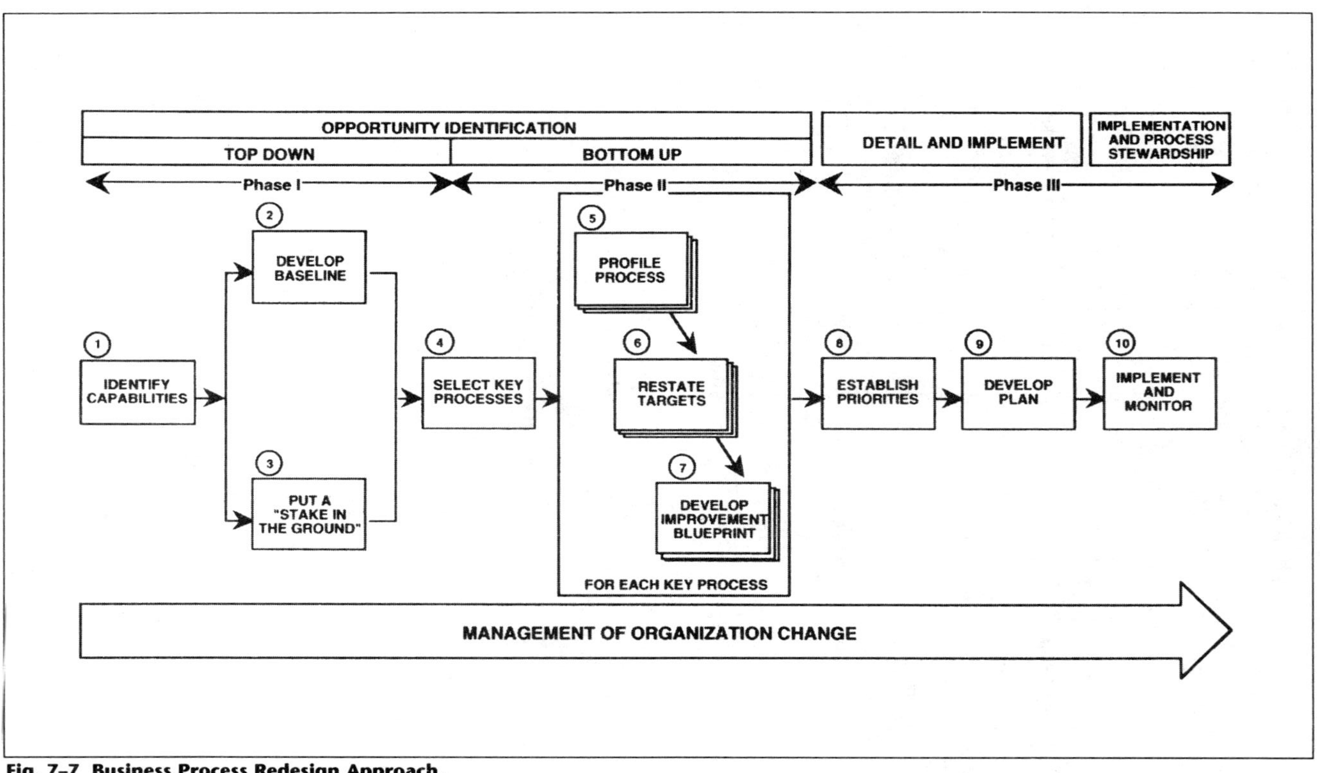

Fig. 7-7 Business Process Redesign Approach.

STEP 2 DEVELOP A BASELINE

The baseline step is triage. We could look at every process from a "produce oil and gas" to a "purchase paper clips." But since BPR can be a significant disruption to the day-to-day activity of the organization, it is important to focus on the most critical areas for improvement. Where are our costs out of line? What are our customers saying about our service? How much change is needed to achieve the kind of bottom line performance our shareholders require?

The economic and service baselines are straightforward. Companies are accustomed to the kind of analysis that can be measured in dollars or in customer satisfaction. A thorough BPR baseline will also include an assessment of the climate for change within the organization. New processes will create change. It is important to understand how the organization is likely to react to that change so that implementation can be planned accordingly.

A side note on customer service: We all think of the royal "They" who buy our products and services when we think of customers. There is another important group of customers: the internal customer. How we serve each other in an organization often determines how we will serve our external customers. It also has a large impact on our cost position. Therefore, when developing the BPR baseline, it is necessary to understand who the key internal customers are and how well they are being served by both staff and line organizations.

Because of the magnitude of assets and investments in an oil and gas company, the baseline often focuses on the performance of assets, the congruence of investments with the environment and company strategy, and the services provided by overhead functions. All three of these areas usually yield opportunities for reduction or redirection.

STEP 3 PUT A STAKE IN THE GROUND

There are basically two approaches to reengineering. We will call them the shotgun and the rifle. It will quickly become clear that we advocate the latter.

The shotgun approach takes its name from the wide spray or pattern that a shotgun uses on a target. Shotgun companies decide that BPR is inherently a good tool, and they apply it uniformly to processes, confident that once the process is reengineered, it will perform at the "right" level of cost, time, and quality, where "right" is defined as whatever the final process result is.

In contrast, the rifle approach takes its name from the single rifle shot aimed at the target. Rifle-shot BPR sets a required performance level and then asks the question, "How must our processes perform in order to reach that targeted performance?"

It is far better to send a team out with a charge of "Improve cycle time by 30%" or "Reduce costs by 40%" than "Go redesign and see where we get." This is what we mean by a stake in the ground. Top management sets an expectation for the redesign teams that they can use continuously to evaluate their progress.

This stake can be set two ways. Analysis could be used to target a cost reduction that will provide required bottom-line performance. Alternatively, management can set a stake by fiat, using it as a rallying cry for the company.

Either way, the stake should be aggressive—there is no point in launching a BPR effort based on only incremental change. This aggressive stance may cause some pain, but the pain may in turn unfreeze people's minds and lead them to step-change improvements.

In the site selection example mentioned previously, management, by proclamation, set a cost improvement of 50% and a cycle time reduction of two-thirds as the goal. This was done before any benchmarks were obtained to understand if this was even feasible.

STEP 4 IDENTIFY PRIME BUSINESS PROCESSES

What is a business process, anyway? Let's look at its component parts. A business process has:

■ ***A customer.*** This is a good place to start. Identify both internal and external customers.

■ ***A product or service.*** A business process has an output. Broadly defined, this can include real products, information, decisions—anything we provide for our customers.

■ ***Process participants.*** These include not only the people in your organization who actually make the products and services, but also their lines of command. A broad analysis of process participants can often be surprising. For example, in the service station example, the company discovered 100 people were involved in the site selection process.

■ ***Process enablers.*** In addition to people, there are systems, assets, purchased services, forms, information flows, and so forth.

If we use these elements to define processes and recall that processes will cut across organization boundaries, it is possible to develop a complete list of the processes being performed in your organization. It will also be possible to allocate all people and activities to one or more processes. (Hint: If you find any people that are involved in no processes, you have found your first "early win.")

With this inventory in hand, this step identifies the most important processes to carry into reengineering. Priorities should be set based on our initial look at capabilities (Which processes are most important to delivering needed capabilities?) and our stake in the ground (Which processes will get me to needed performance levels?). There is no cookbook. Senior management's judgment is needed to balance potential improvement with an understanding of how much change the organization can assimilate at once.

In our site selection example, the prime business processes turned out to be real estate transactions, obtaining permits, legal processes, and decision making. Notice that deciding where a process begins and ends can be somewhat arbitrary. In the end, defining where the process ends is a question of deciding how much a BPR team can bite off at once.

STEP 5 PROFILE EACH PRIME PROCESS

Just as the baseline created an understanding of overall organization performance, this step develops an understanding of prime process performance. As one client said, processes should be "good, fast, and cheap." In this step, we analyze just how good, fast, and cheap they currently are.

The process profiles include both quantitative and qualitative aspects:

- Process maps or flowcharts that describe participants, customers, activity steps, timing, information flows, systems used. They can be as simple or complex as required. An example is shown in Figure 7–8.
- Process economics to describe both the cost of the individual process steps and the overall process
- An understanding of how the process contributes to overall organization performance
- Process headcount, measured in full time equivalents (FTEs)
- Problems with today's performance as raised by process participants or customers

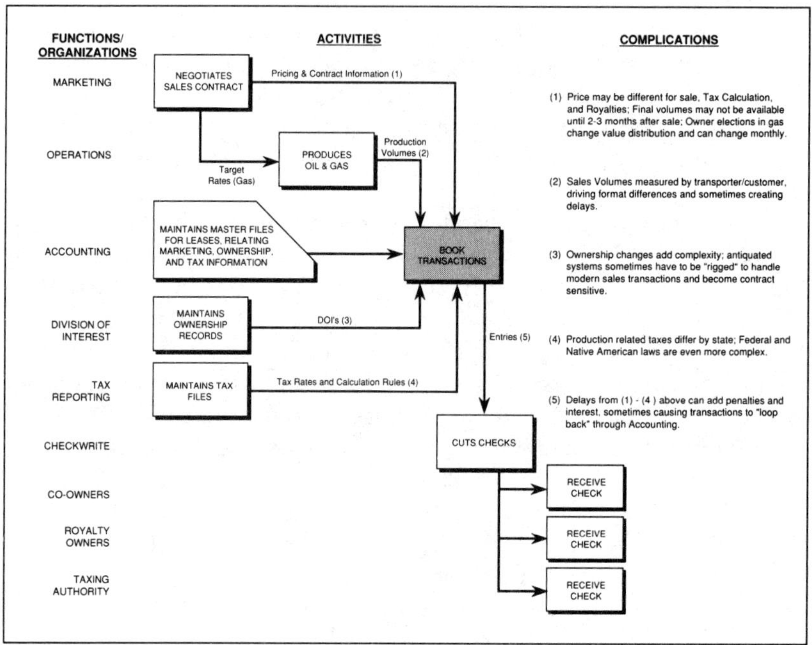

Fig. 7–8 Process Map Example.

- Process timing, both elapsed time and actual activity time, for both individual steps and the overall process
- Process variations (e.g., oil vs. gas, or self-operated vs. outside operated)
- Process products and services
- An analysis of how information technology is used in the process
- Current process performance measures (if any exist) and levels of performance

This step often provides the "Aha!" that spurs top management to action. Because process profiling is not a standard performance analysis tool, management is often unaware of exactly what is going on. In our site selection example, no one had ever asked the question, "How much are we spending per site?" or "How long does it take to obtain a site?" Presented with such analysis, the impetus for change is clear and compelling.

A side note on activity-based costing (ABC): ABC has been both a powerful tool for BPR and a popular capability in its own right. Simply

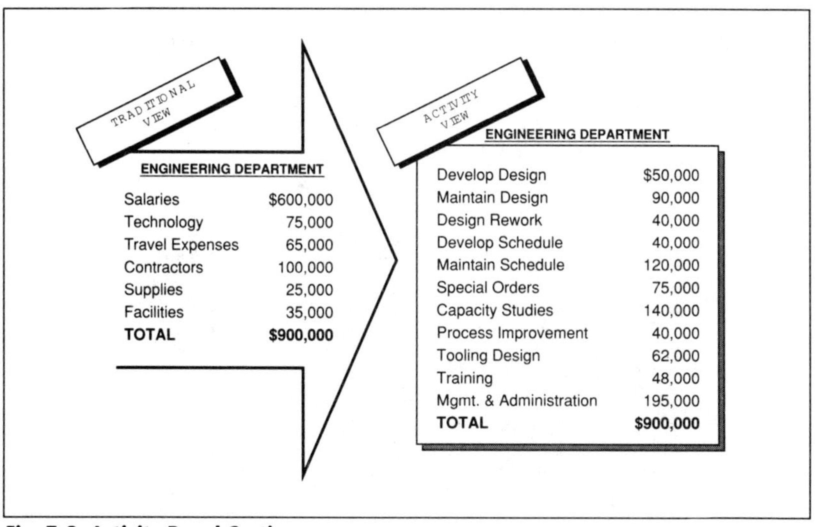

Fig. 7–9 Activity Based Costing.

put, ABC allows us to cost activities instead of functions. Figure 7–9 shows the difference.

ABC is useful in process profiling. The two go hand-in-hand: Process profiling helps to define the activities that would go into an ABC system. ABC then develops the costs for process activities. Left in place after BPR, ABC is useful for measuring ongoing process performance, which of course facilitates continuous improvement.

STEP 6 RESTATE THE TARGETS

Armed with a thorough understanding of the major processes, the team should then revisit the original stake in the ground, restating the targets for each specific process. The team should be challenged to stretch their thinking, to come up with ways of reducing the costs and/or time involved in the processes by 20, 40 or even 60%. Do not be overly constrained by the challenges on implementability.

Also make sure that the team looks first for places where processes can be eliminated entirely. One of our favorite examples involves a client team that was examining a process best described as "the weekly estimate of the monthly budget variance." The team set about diligently to redesign the process, looking at hand-offs and information technology. The better question was why this process was needed at all. A quick examination of the

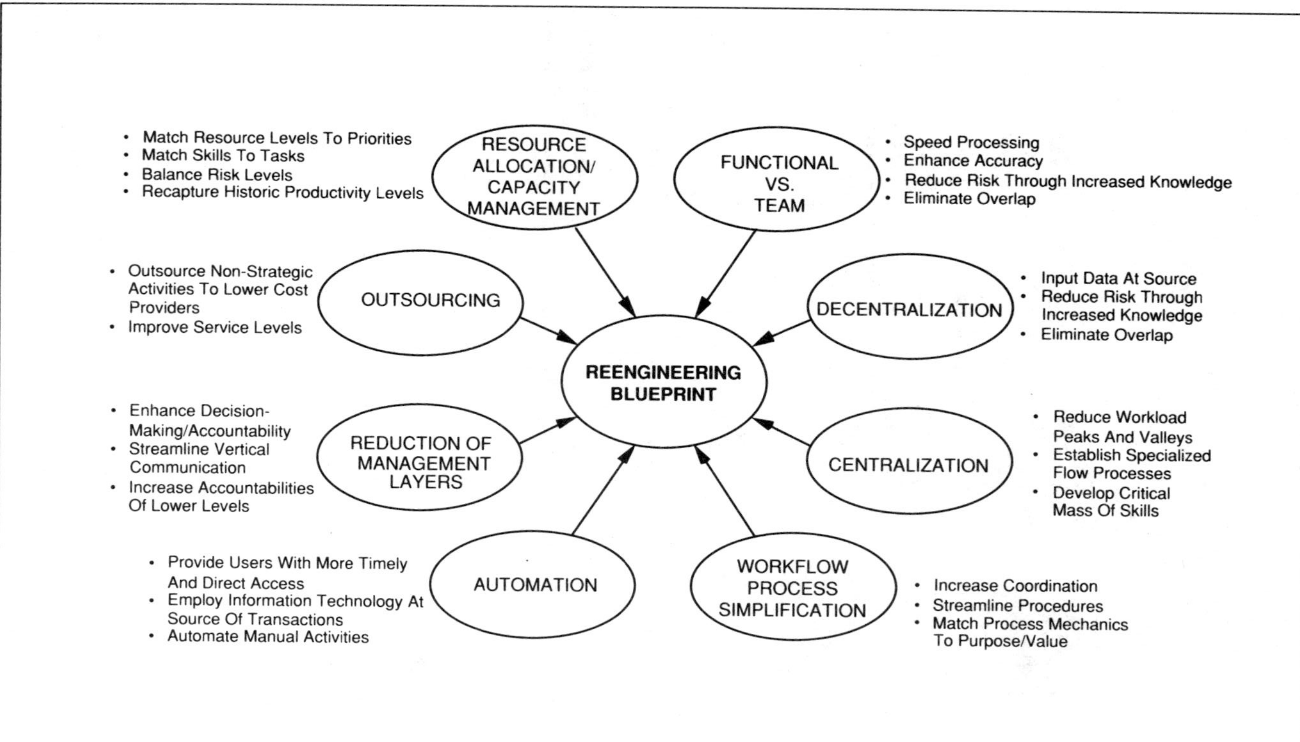

Fig. 7–10 Process Redesign Tools.

Blueprints should be composed of 15 components

- Executive summary
- Objectives, key success factors and capabilities
- Metrics and targets
- Blueprint options, alternatives and scenarios
- Blueprint 'walk-thru'
- Process and information flows
- Job design
- Organization
- Technology, equipment and information requirements
- Financial analysis
- Key blueprint dependencies
- Gap analysis
- Key implementation initiatives
- Transition plan
- Appendix of required supporting details

Fig. 7–11 Process Blueprints.

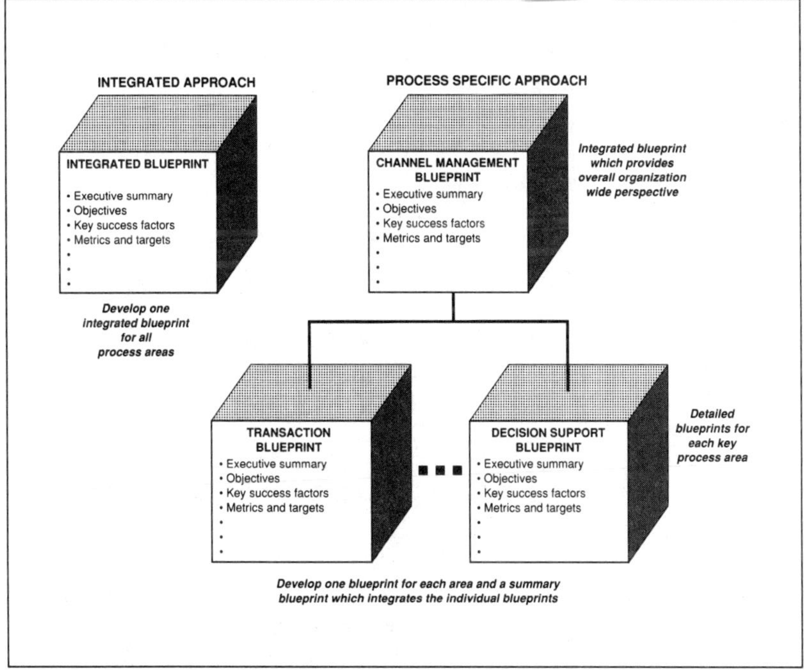

Fig. 7–12 Blueprinting Approaches.

decisions made based on the process (none) and the quality of the estimate (low) convinced the team to eliminate the process rather than waste time redesigning it.

STEP 7 DEVELOP THE BLUEPRINT

Now the real redesign can begin. Using the tools identified in figure 7–10, the team should begin to develop the blueprints of how the processes will work in the future to meet the restated targets. Since there is no cookbook recipe for creativity, it is important to create an environment within the team that fosters creativity. There is interesting data that suggests that our creativity declines rapidly after the age of five. Watch an infant learn to walk; she walks, she falls, she laughs, she gets up and tries it again. In short, small children are not the least bit self-conscious about the mistakes that come with learning.

School changes that. The loss of face that occurs when we answer questions incorrectly in class conditions us to be very careful when we open our mouths. This is difficult for a BPR team, which requires the rapid generation of the occasionally crazy idea. So it is necessary to create the kind of team environment that actually encourages team members to generate new ideas. Some of the most effective BPR teams we have seen look to the outsider like anarchy.

The blueprint needs to be a complete description of the future. Figures 7–11 and 7–12 show the necessary components.

Why all this detail? The blueprint must create a compelling case for change. Also, the blueprint will be the owner's manual for the new process; those responsible for implementation will need the detail it provides.

Information technology (IT) is an especially important component of the blueprint. The new design should include those IT capabilities that will enable improved process performance.

STEP 8 SET PRIORITIES FOR CHANGE

The final three steps represent the actual implementation of the redesigned processes. First, the team needs to understand all the implications of the redesigned processes and decide what gets implemented in what order.

The process redesign will drive many changes. For example:

- How will the redesign affect organizational structure?
- What resources—people, skills, funding, assets, IT—will be required?
- What performance measures will be used?

This step also helps decide what to do next. Often, a BPR team will generate literally hundreds of opportunities, often creating much more change than a company can assimilate at once. It is necessary to decide what the most important opportunities are and pursue them first.

The important lesson to take away from successful BPR efforts is to set a manageable agenda for change that captures all opportunities in reasonable chunks.

STEP 9 FORMALIZE AND INITIATE THE REDESIGN PLAN

The business process redesign plan gets very specific. It begins with the blueprint and assigns detailed action steps to every aspect. It specifies the who, what, when, where, and how of:

- Communicating with everyone who will be affected—both inside and outside the company
- Training participants to perform the redesigned process activities
- Establishing schedules
- Setting implementation performance measures
- Staffing to meet the new process requirements
- Deploying information technology to support the new processes
- Preparing budgets
- Diagnosing the organization's likely reaction to change
- Establishing contingency plans
- Ensuring ongoing stewardship of the process

This plan should be owned by senior management, and it should be highly visible to the organization. At this stage in a major BPR initiative, all stakeholders will be aware of the change but potentially ignorant of the details. Good communication of implementation plans can significantly reduce fear and accelerate the process.

STEP 9 MONITOR THE REDESIGNED PROCESSES

At this point, it is tempting to say, "Just do it." In fact, the greatest challenge in process redesign is making sure it keeps on working. The business process landscape is littered with the remains of good ideas that went nowhere. Many of the casualties can be traced to the horizontal nature of redesign—its effects are felt throughout the company, and any functional department that chooses not to play can effectively kill a new process.

Obviously, strong support from the top of the organization is required to send the message that there is a corporate commitment to the blueprint for change. In addition, we recommend the creation of a process steward—a line manager in one of the organizations that participates in the process. He or she will be responsible for process implementation, ongoing process performance measurement, and continuous improvement. We will discuss this topic more at the end of the chapter.

This raises the issue of the relationship between BPR and Total Quality Management (TQM). Companies often ask if the two are compatible. Specifically, a company that has installed a TQM culture and organization often views BPR as a threat. "Why do we need BPR," they ask, "when TQM is delivering results?"

In fact, the two are mutually supportive, as shown in Figure 7–13.

The key distinction is in magnitude: BPR provides step-level change on a broad scale, while TQM provides incremental, continuous improvement on a lesser scale.

Perhaps a few more examples will make clear the power of this methodology in delivering significant performance improvement.

EXAMPLE 2 PROCUREMENT AND INVENTORY MANAGEMENT

Materials management is a key area of concern for an Asian operating venture of a major multinational energy company. Purchased materials, primarily for drilling and completion projects and production operations, represent about 60% of total operating expenses. Furthermore, in this exploration and production business, the danger of not having mate-

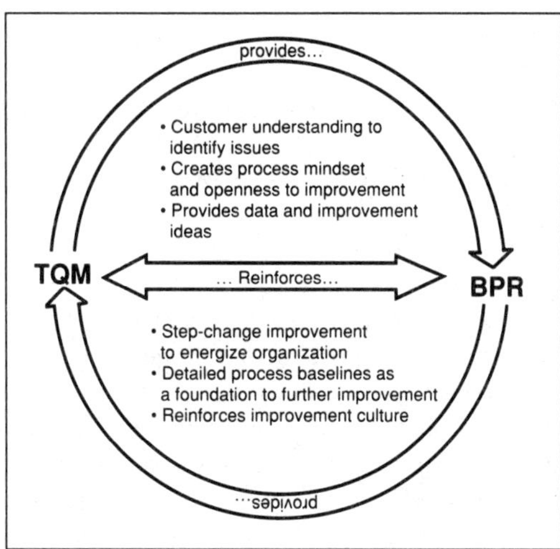

Fig. 7–13 BPR and TQM.

rials when needed poses severe risks to safety, not to mention operating efficiency and revenue.

The company was experiencing difficulties in its materials management effectiveness. Inventories were high and growing rapidly. Service levels were poor and were beginning to affect operational efficiency. Process times were long and erratic: normal lead times for placing orders were over 100 days, and many purchasing requisitions were outstanding for over a year before orders were placed, if they were placed at all. This situation was complicated further by the operating environment. Most field activities were in a remote area of a developing country. Infrastructures, including telecommunications, were substandard. Supply chains were long and complex. Heavy reliance on local staff created difficulties due to both language and the shortage of key skills.

The company launched a program to redesign the overall materials management program. Highlights of the redesigned process included:

■ ***Organizational Integration.*** Materials management responsibilities were originally diffused among the maintenance, engineering, operating, and warehousing organizations. Furthermore, each of these organizations operated independently, and purchasing also reporting to another independent organization.

The company created a new inventory control group to coordinate and manage materials in the field operations. This group was then integrated directly with the purchasing organization along commodity lines.

■ ***Streamlined Operating Procedures.*** Previous procedures were complex and often inconsistent—an amalgam of outdated company policies, local regulations, and specific corrective procedures put in place to correct past problems. All transactions flowed through the same process, with little consideration of types of materials, usage patterns, or procurement issues.

The company implemented a segmented, streamlined set of materials management and purchasing procedures, tailored to each particular set of requirements for different commodities and situations. They set inventory control parameters and procedures to optimize inventory and service levels based on the requirements of each major control segment. Specific programs for order consolidation, family buying, and project coordination streamlined the purchasing process. New vendor performance tracking and screening procedures made purchasing more proactive. Finally, new training programs, reference materials, and information systems tools helped support and institutionalize these new procedures.

■ ***Realignment of Responsibilities.*** The organizational restructuring provided the foundation for new roles and responsibilities. However, in order to fully realize the benefits of the new organization, the company needed to decentralize both authority and decision making to allow the organization to respond more quickly and efficiently. Previous approval schedules were long and arduous, requiring many levels of signatures, and often requiring new, unique documents to be prepared for key steps along the way.

They radically restructured approval levels, eliminating many authorization steps, and decentralizing most routine decisions down to the working team level. Extended authorizations were reserved for large, complex projects, where external approvals were generally required due to local regulations. Standardized forms simplified remaining approval steps, leveraging information systems capabilities to further streamline procedures and reduce delays.

Results were achieved rapidly and dramatically. Inventory turned around, dropping by over 15% in the implementation period alone. At the same time, service levels improved, with stockouts declining 40%. Procurement lead times dropped by almost 60%, while enhanced controls resulted in improved ratings in the corporate audit. Operating efficiency improved, and costs declined. The organizational changes were accomplished with a net decrease in staff.

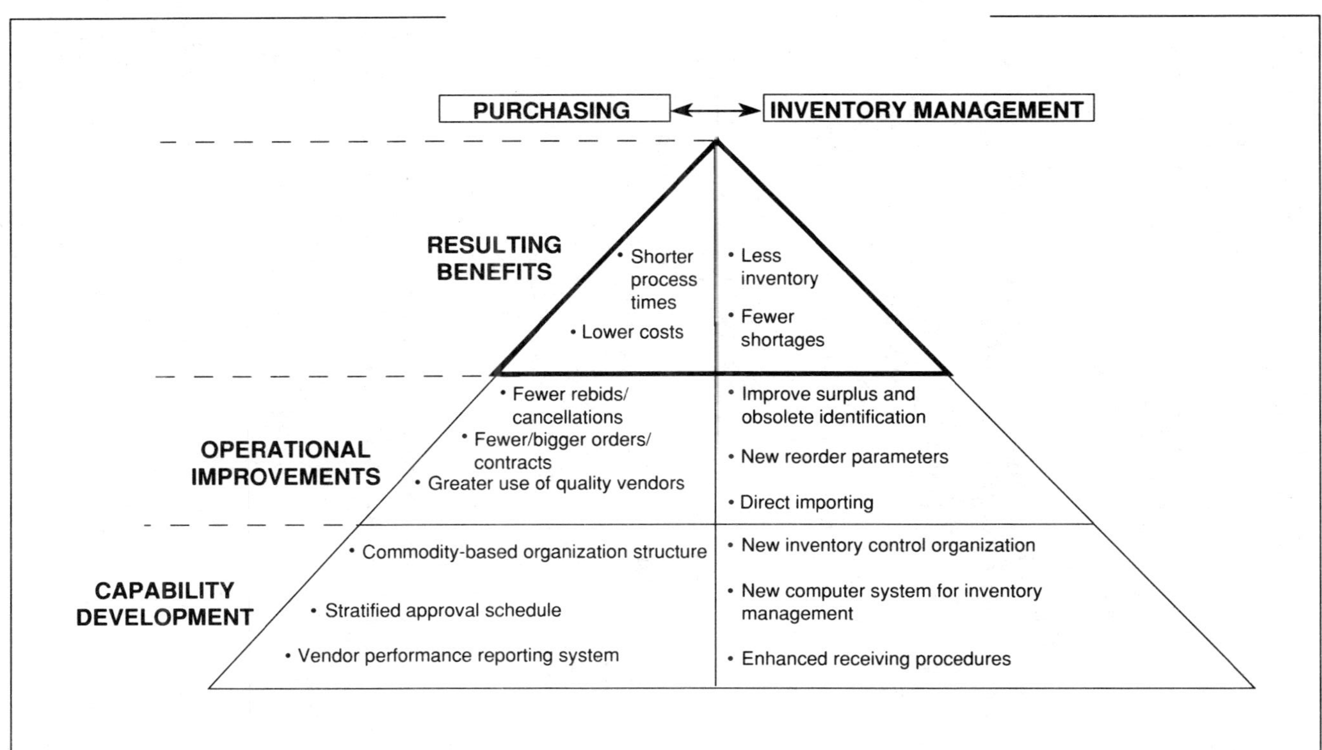

Fig. 7–14 Materials Management Programs.

The program paid for itself well within the implementation period, and ongoing annual cash savings at the time of the postimplementation reviews were estimated to be over $5 million per year. Most significantly, post-implementation reviews one and two years after implementation indicated that the BPR changes had been fully institutionalized and that performance had continued to improve, even with changes in management in several of the areas involved.

Figure 7–14 shows the overall program.

EXAMPLE 3 SITE ENVIRONMENTAL REMEDIATION

Not all redesign results in the direct reduction of staff. In fact, in the following case, staffing levels were increased so that an overall controlled cost reduction in excess of 30% could be achieved.

This example involves the costs associated with the testing, planning, approval and remediation of soil and ground water under an oil company's service stations.

For many years, the gasoline tanks and associated piping underground at a service station have been made of single-walled steel. The need for protecting barriers around the tanks and lines was not even realized until it was determined that benzene (formerly a major constituent in gasoline) was a carcinogen. Then in 1974 the Safe Drinking Water Act was passed (with amendments in 1986), establishing a timetable for all underground tanks to be replaced with systems incorporating additional protections against gradual or traumatic leaking.

The oil companies began the ramp-up process to change over their entire network of stations—2,000–3,000 stations in many cases. As with any rapid ramp-up, the resulting processes and practices were not particularly efficient or effective. By the time the need for redesign was identified, many companies were spending in excess of $100 million per year on this process (excluding the capital cost associated with new tanks, lines, and monitoring systems).

One first step toward BPR was to develop an improved understanding of the costs and their drivers through process profiling. This highlighted the magnitude of future costs, exposing the fact that total costs would continue rising to well in excess of the current period's if strong action were not taken. This created the beginnings of a shared feeling for the need to change.

In parallel with this effort, benchmarking was carried out to identify best practices through the industry. As is often the case with benchmarking, no one company incorporated all the best practices; rather, lessons were learned from each of the companies contacted.

It was becoming clear that costs were rising out of control and that the remediation process was not producing the best results despite all the money being spent:

- Externally, bid jobs were awarded on an inconsistent basis—usually not cost
- Unnecessary work occurred at many sites, while at others work was falling through the cracks because of the high workloads on the in-house staff
- Purchasing strength was not used to its full advantage

Changes in the process and organization focused the department on the critical aspects of the process. Many of the engineers in the department moved to the field. This decentralized organization was more able to focus on the critical aspects of the remediation process—the correct assessment, development, and approval of the remediation plan for a site.

Much of the engineering work associated with remediation was carried out by external environmental engineering firms. While these firms clearly were the most capable of completing the task, the contract terms were cost plus and therefore, their incentives were not aligned appropriately with the company's. New contract terms sharply increased the fixed cost portion of the contracts, resulting in better alignment of incentives to manage costs.

In other areas, such as the purchase of analytical lab services and soil disposal, there was a large benefit associated with centralized contracting. Consolidation of the lab activity reduced lab costs by up to 40% by both reducing costs at the labs (e.g., centralized billing, standardized reporting, and higher average utilization) and decreasing prices to be more in line with costs. Accounting and control tasks were reallocated and redesigned so that the accounting department could take more responsibility for the paperwork, freeing up the engineers in the department to work on the remediation process rather than bill tracking.

Finally, to ensure that the identified cost savings were not lost, a monitoring system was established which focused on the key cost drivers. This was developed on a low cost basis using information that was currently available or could be easily provided by the outside contractors (e.g., environmental engineering firms or analytical labs).

The blueprint for change incorporated the learning from the internal and external reviews. Since other processes were going through BPR, the company had to set priorities among many opportunities for improvement. The priorities in this case were based on both ease of implementation and the magnitude of expected savings. Thus, while services such as analytical labs were consolidated early on because that was easy to accomplish, the consolidation of environmental engineering services was to take place over a much longer period.

It is interesting to note that the changes in the department, the goals, the incentives, and the processes resulted in all parts of the organization being more excited about their jobs and what they could accomplish in getting the job done. What had been a department staffed by people frustrated with a sense of "our hands are completely tied" became one where innovation and contribution was common.

EXAMPLE 4 VALUE CREATION IN EXPLORATION AND PRODUCTION

A leading integrated energy company with a significant E&P business designed a new value development process that addressed value creation from all opportunities and assets. The scope included everything from exploration plays through project development, into production and eventual field abandonment.

The "traditional" approach to the upstream business has strong functional disciplines leading to a fragmented process comprising exploration, development, production, and abandonment. Typically, this creates multiple hand-offs throughout the field life cycle, lack of ownership, and turf battles for the allocation of both human and financial capital.

The new process enabled opportunities and assets to be evaluated as a global portfolio with the primary aim to maximize value creation by applying a consistent evaluation and ranking methodology in order to optimally allocate scarce resources (human and financial).

The new process was based on the company's mission, vision, goals and strategy, shown in Figure 7–15.

The aim of the BPR program was to help implement the strategy and to improve performance. Very early on it was clear that maximizing value creation in an environment with limited cash availability mandated greater selectivity of both exploration prospects and development projects on a global basis. The initial process profiling work identified many problems with the current process:

- No consistent process existed to allow "apples to

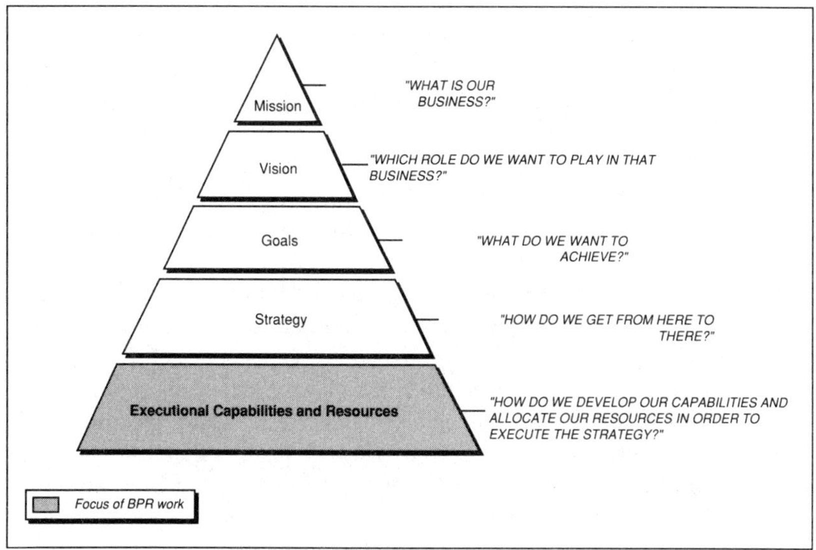

Fig. 7–15 Mission, Vision, Goals, Strategy.

apples" comparisons for investments (exploration or development). Those who shouted loudest got the funding.

- Long development lead times: More than 15 years from discovery to first production, in comparison with industry averages about 10 years.
- Finding costs of \$3.60/bbl in comparison with top quartile performance of \$2.50/bbl.
- Large engineering project teams and adversarial contractual arrangements with engineering contractors, causing slower and more costly development.
- Unwieldy budget cycles involving significant resources over a seven-month period.

During the work, some 24 prime business processes were identified. In order to get the biggest bang for the buck, the company set priorities on the basis of value creation potential, scope for cost reduction, service improvement potential, and ease of implementation. They also set process redesign objectives:

- Create a consistent portfolio management process for implementation within six months.
- Cut development lead time by 50%.

- Achieve top quartile finding costs and reduce operated-drilling costs by 15%–20%.
- Reduce capital expenditures by 20%.
- Create and approve budgets within one month.

In addition to specific cost reduction targets, it became apparent that to optimize value creation it would be necessary to look at the business as a constantly changing portfolio of assets and opportunities on a global basis. This required the development of a fundamentally different process: the value development process. This process is based on the E&P funnel, already described in Chapter 3.

Historically E&P companies have been good at assessing opportunities on an individual, stand-alone basis either based on economic measures (net present value, internal rate of return) or technical evaluations (size of reserves, risk analysis). They have not been good in assessing an opportunity's impact on the total portfolio. The funnel provides a framework to understand how an opportunity contributes to the overall company performance and whether an opportunity in fact maximizes the total portfolio value.

The funnel's effectiveness is based on the principle that all assets and opportunities should be assessed within the context of a total company portfolio and actively managed to maximize value and meet the overall company objectives.

A simplified version of the value development process is shown on Figure 7–16.

The process comprises five main steps. In order to enter the funnel a brief screen is carried out to determine if the opportunity fits the current strategy and is commercially viable.

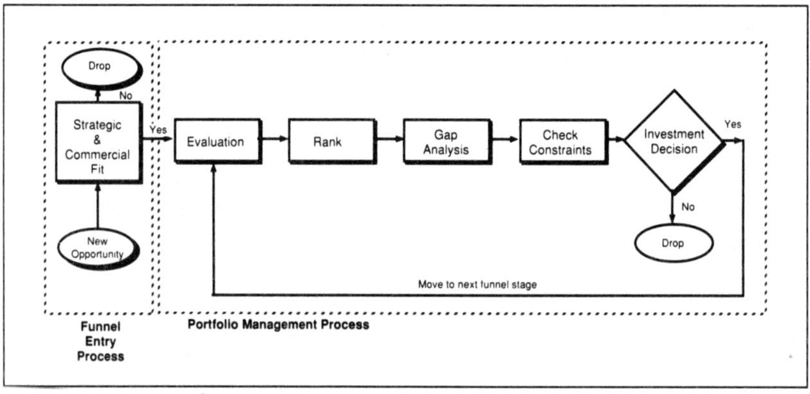

Fig. 7–16 Value Development Process.

Ranking of all opportunities and assets at each stage of the funnel is performed on the basis of chosen economic parameters. A gap analysis determines any shortfalls and surpluses between the ideal company funnel and the current portfolio.

The power of the portfolio management process is that the company evaluates the entire portfolio. If previously determined constraints are exceeded (and cannot be changed), then the total portfolio is assessed to determine how to reschedule to maximize value creation. Finally an investment decision is made to allow the asset to progress to the next funnel stage or a divestment plan is developed.

By successfully implementing the new value development process and funnel concept, the company estimated that approximately $150 million in additional value would be created annually. However, the organization would have to change significantly. Specifically it would have to:

- Move from the traditional E&P approach of exploration, development, and production to adopt the principle of a seamless value development process from exploration through development to production and eventual abandonment (no handoffs, no rework).
- Break down geographical and functional "fiefdoms" by adopting a process to track worldwide performance and create a body responsible for overseeing the processes governing selectivity and allocation of resources (human and capital) on a worldwide basis.
- Create a new culture centered on value creation and a staff that can act in a process-based scheme rather than a functional scheme.

These examples point out not only the broad applicability of the approach, but also the variety of tools that can be used as part of BPR to achieve different objectives.

BPR—WHAT'S NEXT?

We have demonstrated that BPR is a powerful weapon in the manager's arsenal for improving performance. Especially useful for the cost reduction required by the 1980s and 1990s price environment, it has also been used to develop capabilities such as the E&P funnel described above.

Its value has been in giving managers new insight into their businesses. The functional organization is difficult to get a handle on—the kinds of data that come from the vertical silos are not useful in understanding how work really gets done or the attendant performance in cost, time, and value. Process profiles and blueprints have allowed managers to focus on what's truly important.

And it's not just the two-dimensional process blueprints that have added value. The best companies have made the organizations congruent with new processes and a process mindset. This has several aspects to it—structure, performance measurement, job descriptions, and skills, for example.

These organizational aspects are also congruent with the cultures that many companies are trying to install. Consider many of the cultural elements that companies are trying to introduce. Specifically they are trying to be:

- ***Customer focused.*** This is made easier by process understanding, because all processes end with the customer, and the company gets a view of all the organizations involved in customer service.
- ***Empowered.*** Process understanding is critical to good decision making; pushing decisions closer to the line requires giving newly empowered employees the right process performance information.
- ***"Lean and Fast."*** One of the key benefits of process redesign is the potential improvement in speed. Coupled with empowerment, this allows the building of organizations with fewer layers that move faster.

Culture, structure, mindset—all of these are critical to using BPR fully. One key concept in all of this is the idea of the process steward, introduced earlier. Figure 7–17 gives a simple depiction.

In our view, the process steward is critical to achieving successful implementation of redesigned processes. The steward should not be a temporary role; in fact, it represents the solution to the horizontal/vertical tradeoff that the matrix organization unsuccessfully addressed.

The process steward does not own the resources involved in the process; rather, his or her job is to monitor process performance for two reasons:

- To raise a red flag if process performance does not meet the targets set forth in the redesign
- To bring process participants together as appropriate to find further improvement opportunities using the tools of BPR or TQM

The organization of the future then may use BPR to strike the right balance between functions and processes. It will take advantage of the spe-

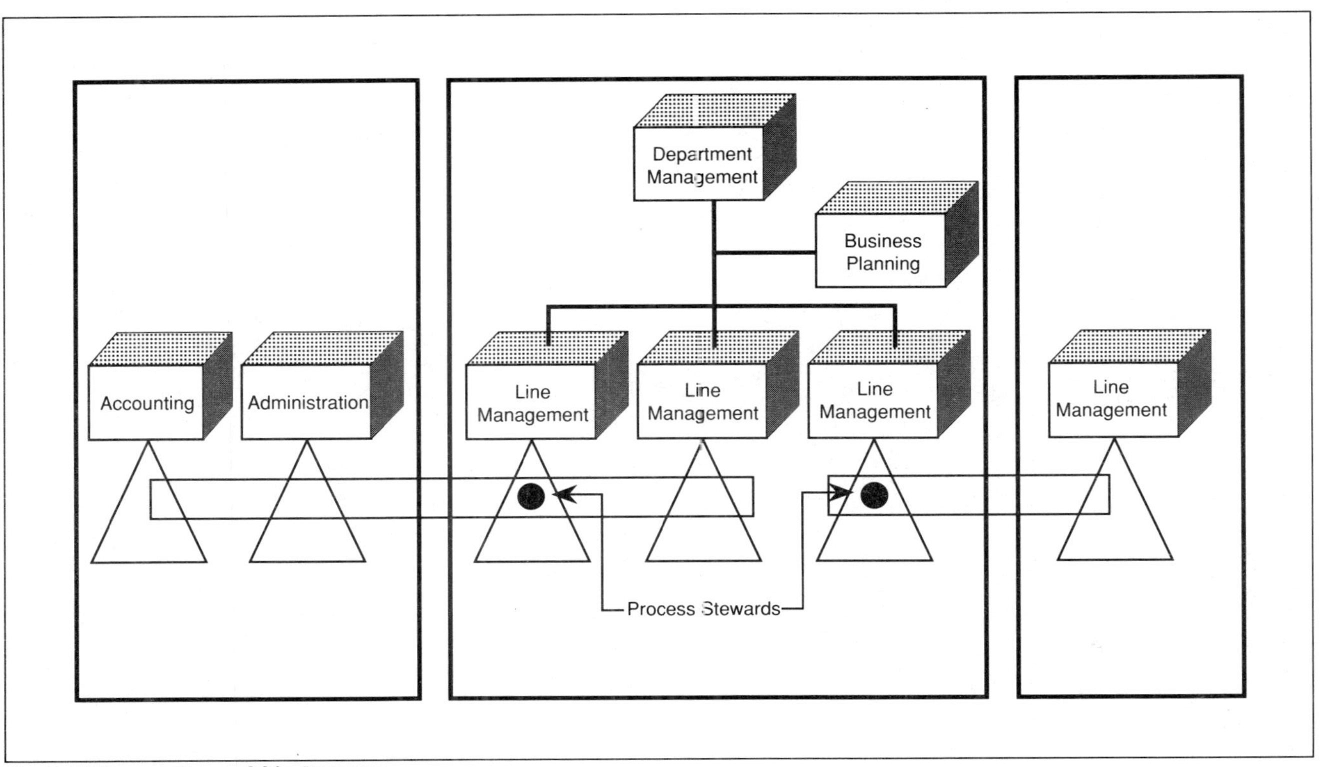

Fig. 7–17 Process Stewardship Structure.

cialization and scale available in functions, while using process stewards to keep process understanding at the top of the agenda.

In the next chapter, we will discuss how these concepts are being translated into organizational designs at oil and gas companies.

---

1 Michael Hammer and James Champy, *Reengineering the Corporation: A Manifesto for Business Revolution*, New York: Harper Business, A Division of HarperCollins Publishers, 1993, p.5.)

CHAPTER 8

ORGANIZATIONAL DESIGN

"When in doubt, reorganize." Organizational change has been such a constant in the oil industry over the last 25 years that many have come to view this phrase as a managerial mantra. One cartoonist suggested major company organizational charts changed so often and unpredictably that they could serve as screen savers rather than strategic enablers.

We believe that a complex corporation's most basic task is to form its activities into natural business units, driven by market requirements. An effective organizational structure provides a sound framework for strategy execution and the flexibility to adapt to a changing market. Every successful business has a few key capabilities that create competitive advantage in its markets—the organizational mission is to develop, to exercise, and to innovate the capabilities its market requires. A successful business has a well-defined culture (way of life) that is focused on these key capabilities. Organizational structure represents an important vehicle for maintaining business culture. A corporation's success depends on effectively deploying capital and human resources against strategic goals. Organizational structure provides a vehicle for capital accountability and career development. The most successful businesses develop broad channels to reach their customers and efficient manufacturing/service processes to remain competitive. The best organizational structures enable companies to strike an effective balance between customer responsiveness and efficient processing scale/scope.

This chapter begins with a brief discussion of the changing business environment. We will then turn to the implications of the business change for organizational structure at both the corporate and business unit level. We will conclude by proposing what constitutes a model of best practices in organizational design and development, looking toward the year 2000.

MARKET UPHEAVAL AS A PRELUDE TO ORGANIZATIONAL UPHEAVAL

The traditional vertical, hierarchical organizational structure served the oil industry well while vertical integration held sway. However, today's rapidly shifting, deintegrated markets demand a new horizontal organizational structure built around markets and business processes.

The large, highly centralized, hierarchical, engineering-oriented organizational structures that characterized the dominant oil companies following World War II derived from clear market fundamentals.

■ ***Scale/Scope.*** Scale and scope mattered. Refinery, distribution, and pipeline investments were characterized by significant economies of scale. Only large players could attract capital and achieve minimum efficient scale production. In this environment, success depends on a few large decisions (when, where, and how much capital to employ). Centralization and hierarchy emerged to ensure effective megaproject decision making.

■ ***Technological Sophistication.*** Each major capital decision exhibited a strong technological component. Engineers, particularly petroleum engineers, became the organizational elite. The corporate culture of most major oil companies became imbued with engineering precision, rigor, and discipline.

■ ***Vertical Integration.*** Vertical integration served to protect these investments from the vagaries of the marketplace. Major oil companies built integrated supply chain production—refinery—marketing to monetize crude. Vertical integration required highly centralized planning—an engineering approach to optimize internal markets. Given the distribution of reserves and markets, vertical integration conflicted with significant geographic focus.

■ ***Multiproduct Production.*** Across the supply chain, oil companies produced several products simultaneously—oil/gas, gasoline/diesel/resid, petrochemicals—preventing a strong product orientation.

■ ***Regulation.*** Oil and petroleum products were highly regulated in many, if not most countries throughout the period, requiring a centralized governmental interface and reducing the value of market flexibility.

■ ***Information Scarcity.*** Information was very expensive to accumulate and communicate. In the era of mainframe computing, data flowed up the chain of command and to the center of the organization.

As a consequence, most major oil company organizational structures reflected this large-scale, technology-driven market.

- Centralized decision making
- Specialized corporate functions
- Hierarchical
- Planning focus
- Rigid corporate structures

Major oil companies were slow but precise, cumbersome but efficient, unimaginative but technically excellent, supply driven instead of customer driven.

Following the OPEC oil embargo, the business environment changed markedly.

Scale Requirements Declined. At post-OPEC price levels, small players could attract capital even without diversification, while technology substantially reduced minimum efficient scale.

■ ***Technology Commoditization.*** Technological know-how became a commodity as innovation drove down costs. Now effective technology application purchased from E&C contractors matters more than proprietary technological know-how.

■ ***Deintegration.*** Markets deintegrated as OPEC merged control of production, exposing most major oil companies to substantial price risk for the first time and emphasizing trading over integration.

■ ***Greater Product Availability.*** The deintegration of the supply chain made individual products available, allowing some companies to specialize in one product category (e.g., gasoline or diesel).

■ ***Deregulation.*** Regulatory restrictions on price fell, allowing significant price competition and placing an emphasis on operating efficiency.

■ ***Information Abundance.*** The cost of information declined rapidly, enabling companies to gather and disseminate more information to more people throughout the company.

This changing business environment places a clear premium on new organizational characteristics:

- Speed
- Flexibility in the face of discontinuities
- Market responsiveness
- Risk management
- Innovation

Achieving these characteristics requires a new corporate culture and organizational structure with the following characteristics:

■ ***Team Focused.*** Building the organization around teams enhances process/market understanding and facilitates rapid change to meet market requirements.

■ ***Integrated Business Processes.*** Core business processes drive market performance. These processes must be integrated from the factory to the customer to eliminate low value-added work and to respond to market requirements.

■ ***Multidisciplinary.*** Teams require multidisciplinary members to succeed. New performance criteria must be established which emphasize breadth of skill as well as depth of skill.

■ ***Flat.*** In the current business environment, hierarchy adds little value; the information and decision rules necessary to win must be embodied in the process and the team. Management then has the responsibility to transfer best practices across teams rather than to make day-to-day operational decisions.

■ ***Decentralized.*** Accountability is being pushed out to move closer to the customer and strategic asset.

■ ***Oriented Around Material Product Flows.*** The supply chain must be tightly integrated to shorten the time to market, minimize working capital and ensure sustained product quality.

Realizing these characteristics in practice requires rebuilding the corporate architecture from the bottom up.

THE ROLE OF THE CORPORATE CENTER

he world's leading energy companies have defined a highly focused role for the corporate center:

■ ***Develop Strategic Vision.*** The corporate center must define which markets to serve, which products/services to offer, and how the company intends to compete.

■ ***Build and Deploy Market-driving Capabilities.*** The corporate center holds the responsibility to identify and to invest in critical corporate capabilities.

■ ***Facilitate the Transfer of Best Practices across the Company.*** As corporate size increases with decentralization, the importance of process stewardship to translate best practices across divisions increases

■ ***Coordinate Relationships with Key Constituents.*** The company's corporate image, political positions, and public presence should align clearly with the corporate strategy and execute consistently around the globe. There are no more local issues and few Antarctic markets.

■ ***Finance the Company Across Business Lines.*** Leading energy corporations top global financial markets, leveraging the company's credit rating across borders and business lines.

■ ***Establish/Enforce Legal/Regulatory Reporting Requirements.*** Market efficiency requires accurate, timely disclosure of financial information—the corporate center holds the responsibility for compliance with the rules governing market efficiency.

■ ***Exercise Fiduciary Oversight.*** The corporate center serves as agents for the shareholders, auditing performance against internal/external policy standards.

■ ***Create a Culture of Accountability.*** The corporate center must substitute strong performance measurement for direct decision making in the new business environment.

■ ***Manage Corporate Risk Exposure.*** Due to conflicting time horizons and incentive structures, individual managers may be taking risks on issues like the environment where the corporation's interests require risk aversion; conversely managers may be risk averse in the markets where the corporation should be taking risks—the corporate center must maintain a balance in the shareholders' interest.

These activities can generally be characterized as fiduciary ownership roles or managerial consistency roles. However, these core holding-company activities account for only a small portion of the typical energy company's corporate staff (see Fig. 8–1).

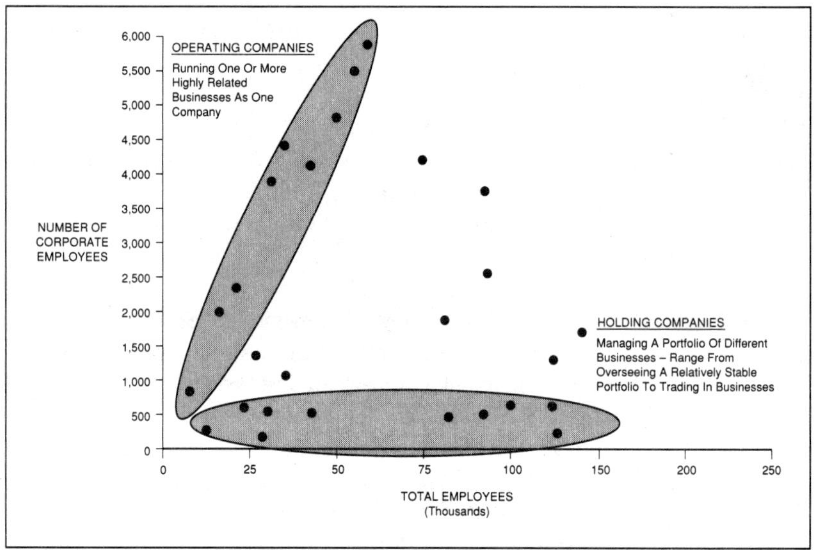

Fig. 8–1 Corporate Level Staff Verses Total Companies' Employees. Source: Booz·Allen Organization Database.

Shared services typically account for the remainder (indeed, a majority) of corporate center expenditure (see Fig. 8–2).

These staff functions—e.g., accounting, purchasing, employee relations, law, information systems—should exhibit strong economies of scale or scope to justify centralization. For example, pooling accounts payable transaction flows across a company may reduce process flow volatility and justify technology investments (e.g., scanning) which individual business units could not afford.

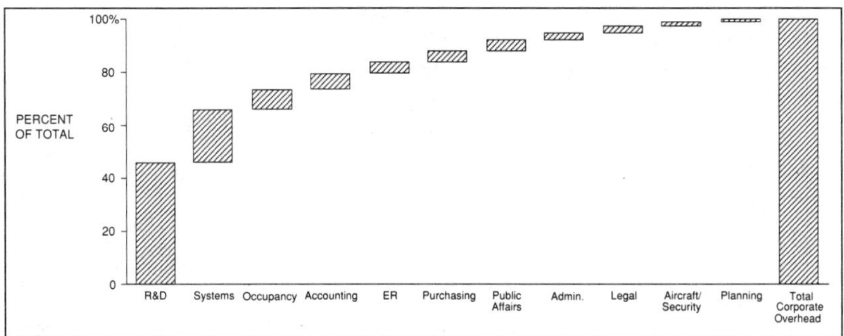

Fig. 8–2 Typical Major Oil Company Staff/Support Activities Corporate Level Expenditures.

The most efficient corporate centers consistently compare shared service performance against outsource alternatives, keeping in-house only the most efficient functions. For example, recently BP outsourced accounting for European E&P and Shell U.S. outsourced information systems application development. Those shared service functions that the company decides to retain then must be held accountable through full cost recovery charge-outs on either a negotiated or a transaction basis. Without adequate charge-out accountability, shared service work loads have a tendency to increase as if the service were free.

The challenge of defining the appropriate role for the corporate center depends on a clear separation of corporate scale for shared services. Then the company must decide what level of corporate roles it can afford and how it can offer shared services most efficiently.

NATURAL BUSINESS UNIT DEFINITION

The most effective companies then invest natural business units with broad authority. These units have the following characteristics.

- Strong management teams (the best people lead operating units rather than corporate staff functions)
- Exceptional market understanding focused on customer relationships and brand value
- Accountable for business results with clear,

results-oriented performance measures and incentives

- Positioned to act quickly, close to the customer with little bureaucracy
- Empowered to act decisively with a wide delegation of decision-making authority and congruence between authority and accountability

The challenge then is to structure appropriate natural business units (NBUs)—entities defined by the market that could be stand-alone companies. Natural business units contain identifiable operations—assets, employees, expenses (not arbitrary allocations). NBUs form around arm's length markets for products and feedstock materials: external suppliers/customers exist; arm's-length prices are set by efficient, competitive markets; industry quality/performance standards govern both intracompany and intercompany transactions. NBUs face free-standing competitors with comparable scope. As a result, NBUs can be held accountable for both income statement and balance sheet performance without significant arbitrary allocation.

NBUs emphasize the most basic form of accountability: the need to earn an adequate return on capital invested. Well-defined NBUs foster rational decision making throughout the organization.

As the external environment evolves, the NBUs must change shape to adapt. In practice, large corporations often permit their organization to lag the market's requirements—masking/blurring marketplace advantages and disadvantages.

Within NBUs, business processes become the primary organizational building block. Effective teams form around reengineered business processes. These teams leverage exceptional local markets to develop advantage market positions. These teams respond quickly to market signals and act decisively to lock in their advantages.

In the upstream, strong regional asset teams have transformed exploration and production decision making. Business processes from exploration site prioritization and drilling through purchasing and accounting need to be integrated (though not necessarily colocated) at the asset team level to produce optimal results. Local knowledge and teamwork deliver powerful reengineering results.

In the downstream, regional, product line, and value chain NBUs provide advantaged market focus and market responsiveness. Two companies have established unique footprints, adopting differing structures across different products and regions. Gulf Coast refineries represent clear, free-standing NBUs. The Gulf Coast has almost 30 world class refineries with access to multiple transportation paths to retail markets. There is a deep spot market for refined products at the refinery gate. Many refineries (such as

Koch, Phibro, Lyondell) have little or no retailing base. Few majors consume more than a tiny fraction of their own output.

Each organizational alternative requires strong business processes, performance measures, and teamwork to be effective (see Fig. 8–3). Companies need a corporate culture that complements and reinforces around the organizational structure.

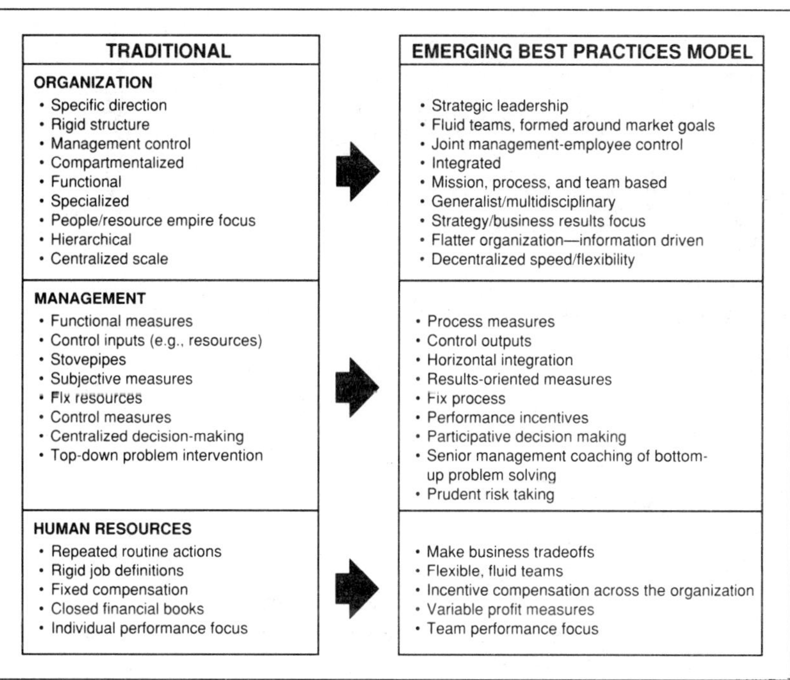

Fig. 8–3 Emerging Best Practices.

This approach also suggests that flexibility and speed to adapt are critical. Organizational structures must be able to adapt quickly to changing market conditions. Yet change is costly. The best organizational structures have developed the capability to change quickly to meet market requirements.

In short, change management is a critical organizational capability. External forces require change. Yet change initiatives frequently fail as a result of internal and external barriers to change. Methods exist to overcome these barriers. The greatest benefits come from institutionalizing change management capabilities (see Fig. 8–4).

In practice, there are multiple hybrid variations on this basic model work—the best answer varies by market. Any one of five themes can be

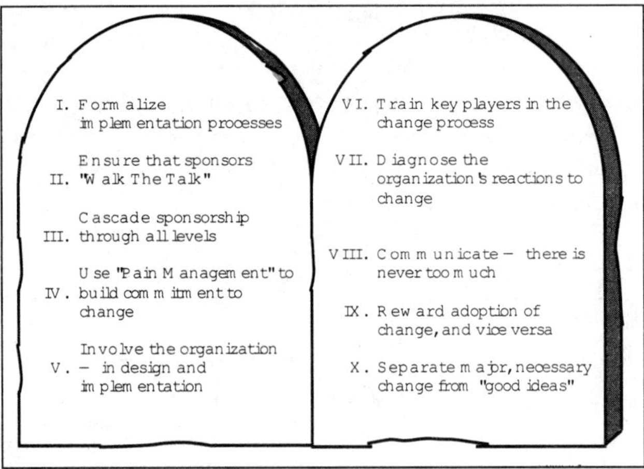

Fig. 8–4

dominant at each level and in each phase of the business:

- Customer
- Product
- Geography
- Function
- Process

For example, a front-end sales organization might be customer focused while a manufacturing organization was product focused. Alternatively, in niche markets an integrated approach built around geographic factors may be most appropriate. Yet sound business processes should form the core of each alternative.

CONCLUSION

We view organizational structure as a fundamental product of corporate strategy. Sound organizational structures, aligned with business strategy and built around reengineered business processes are essential for success. Today flexibility must be institutionalized within organizational structures, allowing teams to respond rapidly to the rapidly changing market. As a result, organizational change will remain a constant in the oil industry for the foreseeable future. Those companies that learn to change frequently and effectively will have built a strategic capability.

CHAPTER 9

THE PLANNING PROCESS

Strategic planning is a term that has been used to describe several activities within a corporation. It has been used to describe the annual operating budgeting process; the annual capital appropriation's process, the capital approval process, and other functions. Let's be clear: strategic planning is none of the above. In its simplest terms, strategic planning is the process whereby corporations define the businesses in which they are competing, as well as how they will compete, and build the requisite resources in order to deliver superior shareholder value. It is externally oriented, observing both marketplace dynamics and current and emerging competition. This definition holds true at both the corporate and operating company level.

This misperception concerning the nature and importance of strategic planning has led many companies astray. Oil companies have not been immune to these missteps. Whether it has been failure to anticipate market dynamics, an ill-fated diversification, or the disappearance of an underperforming company through acquisition, the industry has had its share of embarrassments. Strategic planning, if used properly, can avoid these situations.

The effectiveness of strategic planning can be measured fairly easily. Shareholder returns, as measured by stock price appreciation and dividend yield, provide an objective measurement of the validity of the strategic plan. While it may take several years to allow the market to understand, observe, and reward a new strategic direction, the result can be seen in the underlying stock price. Integrated oil companies have a mixed performance. Viewed over a 20- year horizon, approximately half of the integrated oil companies have outperformed the S&P 500 Index. What will the next decade look like?

Effective strategic planning will determine the future direction of shareholder returns for the integrated oils. With the increasing competitive intensity in each of the businesses, the decline in "elephant" finds of reserves, and the internationalization of the business, strategic planning will be an important tool to direct investments to those businesses which promise adequate shareholder returns.

The intent of this chapter is to examine the evolution of strategic planning within integrated oil companies and to discuss the elements of an effective strategic planning process. The historical context is important in that oil companies are just beginning to realize the increasing importance of strategic planning to their future. We will not explicitly discuss what are effective strategies in this chapter. Those discussions will occur in the chapters for each segment of the business.

EVOLUTION OF STRATEGIC PLANS

he strategic plans, or the quest of making money, of the integrated oils have gone through several distinct eras:

- Vertical integration
- Diversification

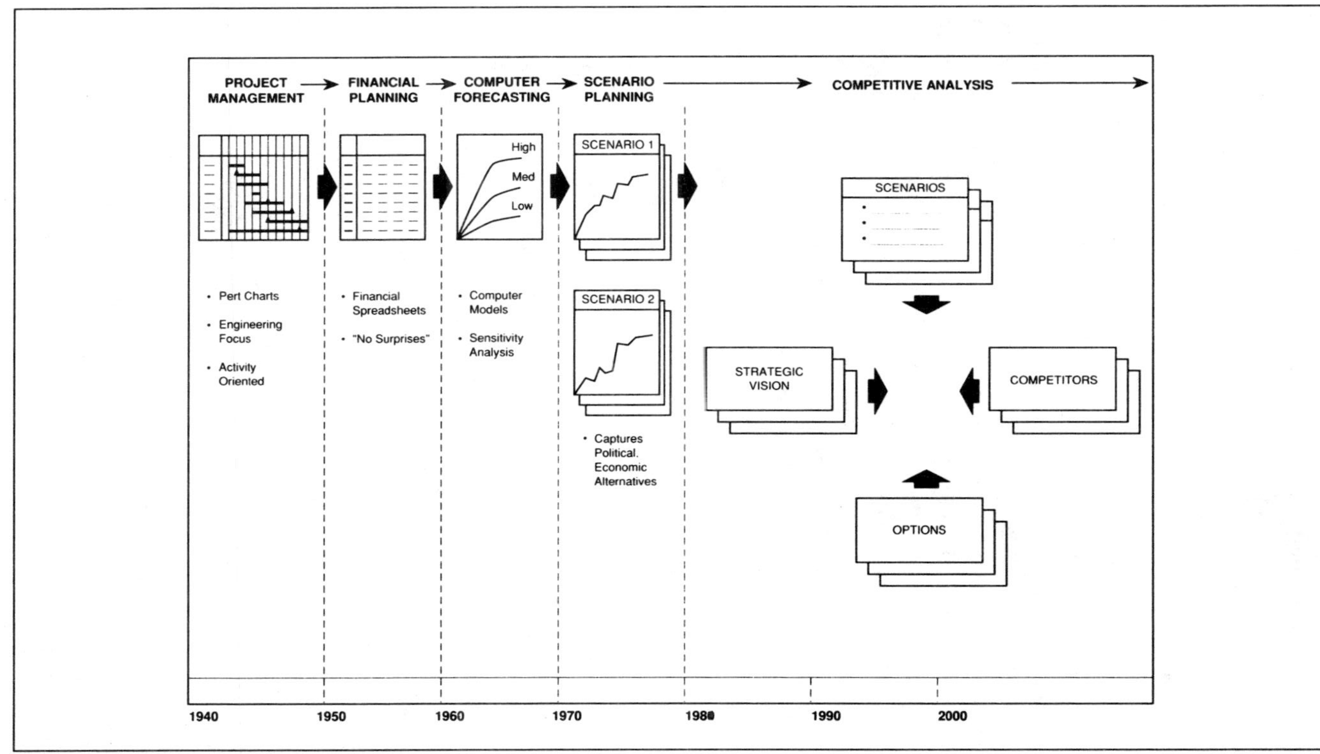

Fig. 9-1 Evolution of Strategic Planning.

- Cost/stand-alone economics
- Internationalization

Evidence of these shifts can be found in several companies annual reports. At the same time, the tools of strategic planning have also evolved. (see Figure 9–1).

For many years, oil companies found that they had to be vertically integrated to make money. The returns, or economic rents, were being made in the upstream business. However, without ready buyers of the crude, oil companies could not realize these rents. Thus came the desire to integrate vertically in order to move the upstream volume. Oil companies built refineries and service stations to serve as outlets for their crude. This practice served an additional benefit in that they could offer foreign countries secure outlets for their crude. Thus came the establishment of Aramco and the volume mentality in the downstream business. Much has already been written on this subject, so further elaboration is unnecessary. Strategic planning was not a major factor during these times for oil companies, nor for the rest of commerce. The competition was nature, not some other company.

As oil prices rose, the free cash flow of oil companies rose commensurably This investment flexibility as well as the fear of scarce reserves started the trend toward diversification in major oil companies. First, they began to think of themselves as energy companies as opposed to hydrocarbon companies. This precipitated the move into shale oil, coal, solar power, fuel cells, and other energy sources. Emboldened by their impressive cash flows, they also moved into unfamiliar businesses. The rationale was similar to the following:

> *The corporation has also been gradually extending its earnings base beyond the energy field into industries where its managerial and technical resources are particularly relevant. (Exxon 1979)*

Witness the acquisition and subsequent divestiture of Vydec, Montgomery Ward, PCA, Reliance Electric, and several coal companies. While it may be hindsight, one wonders if a well-defined strategic planning process could have prevented these mistakes. For example, the decline in the industry was predictable.

As oil prices returned to "normal" during the 1980s, the integrated oil companies found they could no longer earn adequate returns. Thus the emphasis turned to efficiency. However, as the major oil companies were downsizing, streamlining, and so forth, a new breed of competition—the independents in both upstream and downstream—were doing quite well.

The integrated companies failed to recognize changes in their paradigm. First, the emergence of a spot market for both crude and products obviated the need for an integrated oil company. Independents were able to live quite well under the inefficiencies of their larger, integrated competitors. While there may still be managerial value from financial integration (as opposed to shareholder value), there is no longer much value from operational integration. The integrated oil companies are still trying to redefine their businesses and make them profitable on a stand-alone basis. Second, companies were slow to react to the predictable decline in North American exploration. Many companies never recovered from these changes and ended up being acquired. Again, a more robust strategic planning process could have better prepared all these companies for the future.

The future promises more changes. Increasing competition, lack of adequate cash flow, especially for major international projects, declining E&P prospects, and increasing accountability from shareholders will present difficult challenges to the integrated oil companies. Strategic planning will play a very important role in helping companies navigate these challenges and thereby satisfy their shareholders.

STRATEGIC PLANNING TODAY

he modern version of the strategic planning process consists of both recurring and episodic activities.

TYPICAL ACTIVITIES WITHIN THE PLANNING FUNCTION	
EPISODIC	Vision Articulation Business Unit Definition Corporate Strategy Development Development of Performance Measures
RECURRING	Annual Budget Guidance Benchmarking Competitive Position Option and Project Strategic Review Analyses/Capital Expenditure Stewardship Budgeting/Operational Planning Ongoing Performance Measurement

Strategy development should not be an annual event. Once developed, the course should be stayed to determine the performance outcome. This is not to say that progress should not be measured nor that adjustments be made if things are not proceeding to plan. A well-articulated strategy usually has a duration of more than a year, especially in the oil industry where dynamics are not as rapid as say in the software industry. In this chapter, we will discuss primarily the episodic activities.

VISION ARTICULATION

Perhaps the most difficult task for a business or a corporation is the articulation of its vision. A vision statement should set out the reason for an entity's existence. Too often a vision is defined before a strategy is developed. As a result, many visions lack substance. You often hear about corporations aspiring to be the best in their field, delivering superior shareholder value while empowering their work force and respecting the environment. While laudable goals, these are not attention grabbing nor differentiated. If done well, a vision should concisely establish what businesses are to be prosecuted and how.

A good vision is one that is congruent with the strategies of a corporation and/or a business unit. Thus vision articulation cannot be separated from strategy development. They are self-reinforcing. As such, we will spend time first discussing strategy development and then return to vision articulation.

BUSINESS UNIT DEFINITION

A complex corporations' most basic organizational task is to form natural business units. Over the past 50 years, major corporations have responded to complexity by organizing around entities defined by the market that could be stand-alone companies. Stand-alone companies are identified as ones that have identifiable assets/operations/customers. They also have arm's-length transactions with other elements of the corporation. Natural business units impose the most basic form of accountability (the need to earn an adequate return on the invested capital) and foster rational decision making throughout the organization. As the external environment evolves, the natural business units must be redefined to keep up. For example, as the spot market for products has evolved, the refining and marketing business is no longer one business, except in some special instances. The spot market has decoupled refining from marketing so that each must earn a return on a stand alone basis. In practice, large corporations often permit their organizations to lag the external requirements. This frequently masks the competitive dynamics, blurs market signals and places the organization at a disadvantage. A "litmus" test can be used to define natural business units.

The definition of business units is important because it determines the level of granularity for strategy development. Without proper definition, the strategy exercise becomes meaningless. The strategy process should prominently establish the appropriate natural business units.

With properly defined business units, the corporation's role becomes well defined and limited. The corporation's primary job becomes one of setting expectations, building capabilities, monitoring performance, and developing new business opportunities. Thus corporate strategy is distinct from NBU strategy. The following section discusses corporate strategy in more detail.

CORPORATE STRATEGY DEVELOPMENT—NATURAL BUSINESS UNIT STRATEGY

Historically, most companies used an extrapolation approach to planning—look at how they did last year, factor in some changes, and then forecast the ensuing years. This was much more of a budgeting approach, but many companies called it strategy. Needless to say, the results were far from satisfactory. Changes in environmental and competitive dynamics were never considered. By the time the changes were recognized, usually at the bottom line, the company had very little time to react. Dissatisfactions cause most companies to evolve approaches. Today's major companies all tend to use one of the three following approaches outlined in Figure 9–2.

Michael Porter in his book *Competitive Advantage* formalized the premise that a company's objective of profitable growth was limited by external as well as internal factors. This external orientation was necessary in order to define how the business was going to compete. By observing how a business interacted with its environment and an individual company's strengths and weaknesses, a successful strategy could be developed. His strategic concept was that a business could not earn an adequate return unless it established a competitive advantage. He went as far as to identify several generic strategies that could be used to develop competitive advantage. While other strategic concepts have evolved since his book, the competitive advantage context is still the most widely practiced.

CLASSIC (PORTER)	SCENARIO PLANNING (SHELL)	STRATEGIC SIMULATION (BOOZ·ALLEN)
Views businesses being influenced by markets, competitors, customer, suppliers, etc. Rigorous analysis of industry fundamentals Determines ways to achieve competitive advantage Identifies critical success factors	Builds upon classic approach Consideration of various scenarios besides most likely Implication of scenarios Definition of "robust" strategic options/fallback strategies	Expands dynamic aspect of scenario planning Analyzes discontinuities Questions paradigms Integrates approach with process

Fig. 9–2 Strategic Planning Approaches.

The value of Porter's contributions was his emphasis on competitive and other external dynamics—and their quantification. While companies for some time had thought about competitors in strategy development, the analyses of the competition were often limited. And human nature being what it is, competitive realities were often understated.

Oil companies have adopted the competitive advantage paradigm. They recognize that most of their businesses are commodities with little chance of differentiation. Advantaged cost position has become an imperative, especially in more mature domains. The streamlining we have witnessed in the 1990s is a direct result of companies trying to improve their competitive position. Whether it be in production, refining, marketing, chemicals or support services, most oil companies are trying to lower their cost structures.

There have been three limitations to this course for strategy development in oil companies. The first is one of perspective. For the most part, the major oil companies have compared themselves to each other. They have ignored the success of independents. Independents are often dismissed as not real competitors either because their sustainability is questioned or doing business their way is not acceptable. This is shortsighted. In most sectors, independents are defining the competition and are setting the margins available to a participant. The second limitation is the competitive advantage paradigm. Position-based advantages are too transient and can not be relied upon to sustain a business into the future. Building enduring capabilities is the best way to ensure an entity's success for a sustained period. Finally, most companies have based their strategies on a most likely course of events and a static understanding of the competitors. The question is whether the chosen strategy is robust enough to weather departures from the presumed course. Alternative strategy approaches were developed to address this latter point.

Scenario planning was introduced to account for the uncertainty inherent in speculating about the future. It can build upon either the competitive advantage concept or other strategic concepts, such as capabilities. In practice it consists of three steps: (1) formulating scenarios, (2) identifying "robust" strategies, and (3) developing specific strategic action , or changes in strategy, as events unfold.

Perhaps the hardest part of scenario planning is developing alternative scenarios. A scenario is the convenience used to account for the stochastic nature of all the important elements that affect the performance outcome of a business. The scenarios must be sufficiently real to warrant serious debate, but they must be sufficiently thought provoking to question existing paradigms. The complexity is increased by the practical fact that the number of scenarios must be small in order to engage senior management (Monte Carlo simulation is out of the question), and that the scenarios must define all external elements (market, competition, regulation, and so on) in a consistent manner. Nor should the scenarios be so extreme that they have a very small chance of occurring. They should encompass a range of reasonableness (say 70%–80%) but not try to depict either the best or worst case. If the best case evolves, count your blessings. If the worst case evolves, no doubt it will be traumatic, but a business cannot operate expecting the worst case. Considerable time and effort must be invested to develop the right set of scenarios.

Once the scenarios are developed, one tests the existing strategies against each. The likely success of the strategy must be measured against each scenario. As the strategic implications of the scenario are investigated, the existing strategies are modified to optimize the outcome for the business. For each scenario, there is an optimal strategy. However, the larger question is whether one of the strategies is dominant, or robust, across scenarios. Usually, there is not a dominant strategy in the true mathematical sense, but there is usually one that is close. One should always revisit the scenarios if a dominant strategy emerges to assure that they are different enough to explore the range of available strategic options.

As time progresses, the uncertainty in our forecasts is resolved. Thus, as various elements of the forecast unfold, the strategy can be changed, if necessary, to reflect particular events. The last step of scenario planning is the consideration of what strategic actions will be taken if certain key events take place.

Shell pioneered the use of scenario planning in the oil industry. Shell planners claim that this technique enabled them to predict and to prepare for seven of the eight significant changes in their environment since (and including) the first "oil shock" in the early 1980s. In 1988 Shell developed two broad scenarios for the 1990s, scenarios for which the jury is still out.

- ***Sustainable World.*** Triumph of a new global order, environmental concerns are high on the agenda. Collective action/agreements encourage technology transfer; alternative energy technologies grow more rapidly.
- ***Global Mercantilism.*** International order fragments into regional trading blocs an bilateralism; fierce competition among companies and regions; environmental concerns recede in importance. In energy, an emphasis on regional self-sufficiency.

In 1992 the Shell planners reacted to the tumultuous developments on the late 1980s and early 1990s (the break-up of the Soviet empire, widespread market liberalization, and the alarming outbreaks of ethnic and tribal violence from Armenia to Bosnia to South Africa) with two new global scenarios:

- New Frontiers, in which liberalization triggers a sustained period of steady economic expansion
- Barricades, in which ethnic violence, environmental concerns, and economic conflict result in the erection of new trade barriers and a return to autarky.

The development of scenarios can be accomplished by a central planning group, or by a special planning team drawing broad participation from the organization. If managers participate in scenario planning (a time-consuming process taking, in some companies, as much as six months each year), then management buy-in will occur, and those same managers can be expected to take actions that are consistent with the shared understanding of the probable direction the future will take. However if managers do not get involved, the process loses much of its impact. In either case, the process stops short of providing the management team a mechanism for exploring the potential effectiveness of various options for dealing with predicted futures.

When confronting futures with large degrees of freedom, Booz·Allen has also developed an approach we call "discontinuity analysis," in which the team divides into subgroups and develops a wide range of potential "discontinuities" (events that would be sufficiently important to make a difference to a company's strategic direction). From analysis of these discontinuities, scenarios can then be constructed in which several discontinuities are combined. While most oil companies give some consideration to alternative events, only few have formally adopted scenario planning into their strategic planning approach.

STRATEGIC SIMULATION

o this point, there are two deficiencies in the traditional strategic planning process that need to be addressed:

Inadequate treatment of the competitors' range of action and response to one's own actions

Inadequate ability to examine the efficacy of various options to deal with future scenario discontinuities

In the emerging world of the 1990s, competition appears to be increasing as political and economic developments reduce barriers and open markets. In this context, companies no longer assume that competitors will stand out. Strategic planning must therefore attempt to analyze potential competitor actions and reactions, such as:

- Pricing actions
- New market entry
- New technology development
- Strategic alliances

Indeed, competitive assessment is much more challenging.

- Competitors may not act in the future as they have in the past.
- Competitor reactions may be unexpected and the impacts of their actions may be counterintuitive.
- The full range of what competitors are capable of doing must therefore be considered.
- The effects of potential competitor actions on the strategy and reactions to the strategy must be addressed.

Booz·Allen has pioneered a new methodology for competitive assessment which, in effect, enables a company's management team to experience the competitive dynamic of their marketplace. This methodology not only provides unique insight into the dynamic possibilities of future competition, but it also provides a mechanism for examining the ramifications of discontinuities in future scenarios, for integrating the products of previous strategic planning steps (literally of all of the strategic planning models discussed earlier taken together), and for team building and consensus building. We call this powerful new planning tool competitive simulation.

A competitive simulation is for industry what a war game is for the military, a mechanism in which company leaders can "game" the future, playing the roles of their competitors and key stakeholders themselves. Figure 9–3 summarizes the competitive simulation process.

The challenge has been to ensure that the "play" is credible—that the participants, who create a "future" together in the process, find the process to involve those variables which they know from their own experience to be critical, and that the results of the process be believed (although, almost always, the results are a surprise to all because of the unexpected effects of competitor and other stakeholder actions).

Fortunately for the mythical businessman quoted in the box, those painful setbacks happened over four hours in the privacy of the company's executive suite. With a group of his colleagues, he was developing a vision of his company's future by playing out its strategic plan in a "war game."

In a war game—or competitive strategic simulation, as it is called in business applications—company's managers organize into teams representing their company, their competitors, their customers, and others who could affect their success or failure in the marketplace. Each team decides its goals and strategies, then sets out to achieve them. The simulation is open ended. No one knows where it will end or what will happen along the way. That will depend on what the teams do. The objective is for each team to try to achieve its goals over the course of play irrespective of what the other teams may do to thwart them.

In a simulation typically taking 2 1/2–5 days, teams must maneuver adroitly in a marketplace which is changing, against active competitors. They probe for weak spots in their competitors' tactics, then try to exploit them. They look for ways to turn situations to their advantage. Opportunities come and go. Government agencies implement policy, international political, and economic conditions evolve, and the public and the company's customers work to serve their own best interests. Fortunes rise and fall with the quality and timing of what the teams decide to do, and the future unfolds before the players' eyes. Often that future is not what the

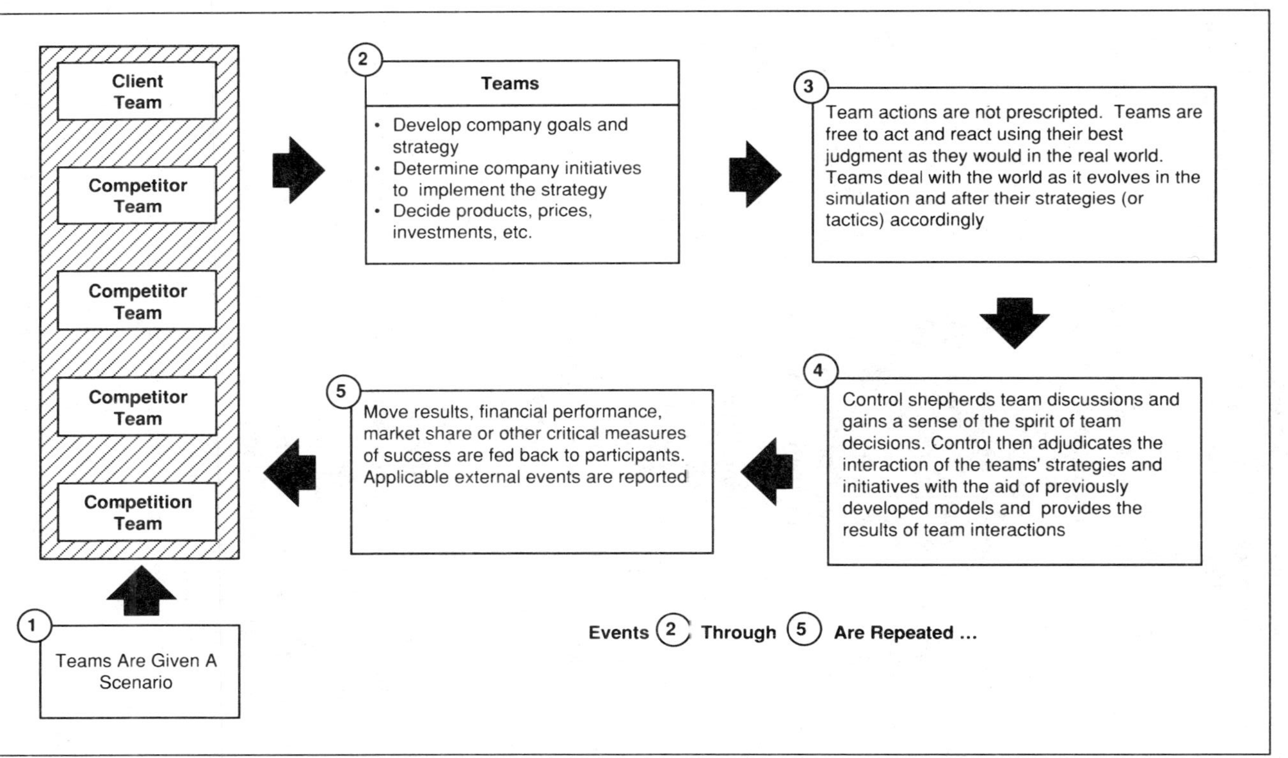

Fig. 9–3 Simulation Structure.

participants envisioned when they developed their strategic plans. Flaws in those plans begin to surface. But all agree that it is far better to be sent back to the drawing board as a result of a war game than after plans are set in motion, when there may be is no turning back.

The military has used gaming for centuries. The U. S. Navy used them to great advantage before World War II. Admiral Nimitz often recounted that the battles of the Pacific had been "fought" during the 1930s on the gaming floor of the Naval War College in Newport, Rhode Island. As a result, he claimed there were few real surprises for the navy as they pushed toward Japan and ultimate victory. In nearly a dozen assignments, the process has proven to be just as valuable for a Fortune 500 company attempting to improve its ability to compete in a changing environment.

Today the prudent CEO, like the prudent military commander, tests plans under the most rigorous and realistic conditions he can find to ensures he has anticipated and planned for whatever the opposition might dish out. The power of gaming as an analytic and assessment tool for business is irrefutable. Simply put, a simulation gives company managers a heretofore unavailable window to experience how their strategic and tactical options might resolve themselves when faced with unrestrained competition. Many times vital discoveries are made about courses of action being contemplated which no participant in the simulation had appreciated before.

"Who knew? I mean, yesterday we were industry stars. We were expanding like mad. We were doing business in two dozen countries. We had a strategic plan calling for revenues to double by the end of the decade.

But the joint ventures in Latin America and Asia fell apart and our cost of raw materials went through the roof. Our two biggest competitors merged and started a price war that got way out of control. Our credit rating was downgraded. We were getting squeezed.

Then we received that not-so-friendly takeover offer. We managed to shake them off, but only after a fight. And then—pow—the accident at our main European plant. You can imagine the mess: government investigations, fines, law suits. Our credibility with customers was close to zero. All that happened this morning. We were just beginning to turn things around when we broke for lunch...."

Currently some well-managed companies do extensive, highly focused, quantitative analysis of key elements of their business plan. Then the best of them do the kind of sophisticated scenario planning pioneered by Royal Dutch Shell more than 15 years ago to position themselves favorably before the coming oil shock. But to stop there is to build a bicycle but not test ride it, not test market it, not see how it stacks up against the competition's response. A simulation is the next and most important step. Because it is in a simulation where the company can actually try out its plans in a realistic environment against a determined competitor.

A simulation gives the manager a way to discover ways to gain important leverage by building bridges between divisions, by giving parts of the company a freer hand, or by using selected strengths to bolster vulnerabilities. A simulation reveals the weaknesses in the company's plans and suggests alternative solutions. The manager begins to see what he could not imagine. "Competitors" may suddenly decide to raise the marketplace ante. Long-standing "customers" may express profound disinterest in the company's new product. "Policy makers" may caste a suspicious eye on traditional business practices. Could changes in operations put a real trump card in marketing's hand? Could a better link between manufacturing and finance create bigger margins? Might some comfortable assumptions be wrong? What land mines lie en route to the marketplace?

Most significantly, however, a simulation lets managers "walk in the shoes" of their competitors and their customers, seeing the world from the other guy's perspective rather than their own. Simulations cannot predict the future. But they can cause the entire management team to experience the forces and dynamics at work in their marketplace. That "experience" has all the attributes of real experience. They have lived through it, internalized its lessons, gained new insights, and can explain it exactly as if it had happened in the real world.

NOT BEING SURPRISED BY SURPRISES

Conventional analytical forecasting typically assumes business conditions will continue into the future much as they have in the past. Unfortunately, linear extrapolations cannot capture what an unpredictable competitor might do, or account for what might motivate a key executive to reach what, from our point of view, might seem to be an irrational decision. Additionally, while we can use conventional analytic techniques to assess a static situation, an evolving situation presents a far more complicated challenge. Over time—as in a game of chess—one tactic opens a range of options and sets into motion a series of moves and countermoves whose end result is impossible to predict. The sheer number of possibilities spinning off even a short series of moves and countermoves is daunting. The farther one moves from the opening gambit, the less initial conditions affect the outcome.

Similarly, future events—like counterintuitive moves by the competition, precipitous regulatory shifts, changes in the economy, political disruptions, or natural disasters—are difficult to assess. Analysts can take them into consideration, however, when they do, the assumptions they must make in doing so are inevitably challenged, along with their subsequent effect on the analytical conclusions reached. The insertion of anything like this into an analysis is virtually impossible to justify. In the end, the analyst is sometimes accused of skewing the analysis to satisfy personal agenda.

This is not the case in a competitive simulation. Surprise events are often included in the design. Participants recognize that the probably of the event occurring is not the point. Instead, the issue is how to deal with it, should it occur? Participants then deal with surprise exactly as they would in real life, that is, according to their experience, the facts at hand, and their understanding of their company's best interests. In this way, any event is a candidate for examination. Introducing it in the simulation will only improve the company's ability to deal with it should it actually occur.

SPECIAL BENEFITS OF A SIMULATION

While a simulation cannot predict the future, a well-designed and executed simulation will:

■ ***Get inside the competitor's thinking.*** We tend to view others through the prism of our own values and experience. Unfortunately, this can blind us to the way our competitors see the world. To best them, we must first see the world as they see it. By assuming the competitor's role in a simulation, a manager begins to see the competitive landscape—and the rationale for both company and competition's positioning—very differently.

Commonly clients question whether their own managers can play the competition credibly. Won't they just mirror image their own company's culture? In fact, our experience is overwhelmingly the opposite. After studying specially prepared briefing books on individual competitors, and after participating in an interactive preparatory session careful coached by the simulation director, company managers quickly assume the corporate mindset of the competitor they are playing. As they do, the reasons for a competitor's particular behavior suddenly become clear. When managers internalize the market position of a competitor, then act in that company's best interests, the resulting play is uncannily real.

Adding to this is a simulation's ability to bring out the competitive reflexes of good corporate managers. After many simulations for many different businesses, we are no longer surprised at how aggressively everyone works to "win."

■ ***Hone implementation skills.*** Conceptualizing a strategy is arguably the easy part. Figuring out how to implement that strategy so it works is quite another problem. Many believe this is the place where most companies are the weakest. To a large degree, good implementation relies on intangibles like intuition, judgment, and a sensitivity to subtleties —in short, the instincts to be able to navigate in uncharted waters. None of these qualities can be taught directly. Each is instilled with the experience of dealing directly with competitive challenge. Consequently, acquiring them in real business situations can take time and can be expensive if gained at the price of bad decisions. A simulation, however, allows a manager to "live" through the consequences of a wide range of actions over the compressed period of the simulation with potential damage limited, at worst, to a bruised ego.

■ ***Gain an appreciation of possible pitfalls.*** Major discontinuities aside, consider the effects of life's little aggravations unfailingly occurring at the least opportune time: a labor dispute, an electrical outage, a change in accounting rules, a recession, new environmental requirements. As the simulation unfolds, managers begin to appreciate that the hinges on which their strategy swings are vulnerable to many kinds of failure. Simulations can make abundantly clear that to keep strategy on track, flexibility and contingency approaches must be built into the plan.

■ ***Build a common strategic vision.*** The most contentious strategic variables typically revolve around cross-functional issues. The product design options that look best to finance, are seen as a recipe for disaster by manufacturing. The operations people want to streamline customer service, while marketing is asking for increased support. Managers tend to optimize activities using criteria that apply to their specific function: what they understand, what they can quantify, what they can justify. Each team in a simulation is intentionally a mix of managers from across the company's functional areas. During the simulation these managers must confront the company fortunes as a whole—often for the first time—and optimize team decisions on the basis of what is best for the company, not simply what is best for their department.

A team's final fortunes are the result of many decisions—some beneficial and some less so from the point of view of any one manager—but arrived at collectively in response to shifting market conditions. Team members share credit and responsibility for the future they create. Not surprisingly, participants sometimes find that their long-held views regarding the company's competitive options shift appreciably. After one simulation, a manager said, "In the end, things were totally different from the strategic vision we had when we went into the game."

Managers who have participated in a simulation tend to carry back to their departments the hard-won lessons that emerged from the three or four days of intensive team interaction. They become team builders in the company. Infused with commonly arrived at and agreed upon company-wide goals and an understanding of the rationale supporting them, they return to all parts of the organization where they inevitably spread the word. The CEO now has managers across the company with a shared understanding of the industry's dynamics and a buy-in to what the company needs to do to compete successfully. For some CEOs, that alone is worth the time to run a simulation.

DESIGNING A SIMULATION

Any business issue can be translated into a game. A simulation can look at a finely focused and limited issue at great depth, or across the whole corporate strategy. It can assess strategic or tactical options, or it can train salesmen specifically to develop and implement account plans in the field. In every case it will improve the participants' ability to make sound business decisions. But to do so, the simulation must carefully crafted to accomplish the client's stated objectives and not stray from those objectives when it is executed.

Whatever the focus, building a simulation is a significant undertaking. Normally it will take from six weeks to three months. Every commercial simulation is the product of a dedicated team bringing together industry expertise from Booz·Allen's national or international commercial practices with simulation and process expertise from Booz·Allen's gaming practice in Arlington, Virginia.

Booz·Allen's commercial practices have great depth of experience consulting to virtually all major industries. Consequently, Booz·Allen brings to each simulation a broad industry perspective plus detailed current knowledge of the client's business. At the same time, Booz·Allen has been the U.S. government's contractor-of-choice to develop and to test new gaming techniques in areas which have proven especially resistant to conventional analysis. This teaming of strong industry and strong simulation expertise is unique.

A simulation's success depends as much on the team's careful preparation of both the simulation and the participants prior to the exercise, as on experienced control of play during the game. Among the preparatory steps are:

1. *Setting the simulation objectives.* This is the corner stone of the project. To ensure the effort is precisely focused on the client's issues—however broad or narrow they may be—the simulation's purpose must be clearly understood and explicitly stated at the outset. As the preparation goes forward and the design team moves from the general to the specific, objectives are revisited with the client regularly to ensure every aspect of the simulation design is aimed at the results desired and the necessary elements of analytic support are built in. It is not unusual for objectives to evolve somewhat during this process, but this inevitably sharpens the focus.

2. *Determining the time frame of "moves."* Each simulation "move" or "turn" represents the time interval normally required for a major decision in the industry to bear fruit. For example, if the implications of developing and rolling out a new product, building a new overseas plant, or forming a strategic alliance take three years on average, then the simulated length of the move would be three years. The actual time for the teams to play that three-year period would be anywhere from three to five hours depending on the other decisions which the teams will have to make in the move. An entire simulation exercise comprises two to three such moves. Time frames are flexible and depend on the simulation's objectives and the time available to devote to it.

3. *Assessing the competition.* This is done prior to the simulation, principally by the Booz·Allen industry experts in teamwork with the client to provide the participants with the information they need to play each company credibly.

4. *Building a top-level industry model.* Working with the commercial team members, a simulation modeling specialist develops a model to capture the intermediate to long-term essence of the business being simulated. The model generates simplified profit and loss statements and calculates share for each player team.

5. *Assigning teams.* We work closely with the client to ensure the teams have cross-functional expertise and a good knowledge of the competition. Sometimes individuals are chosen for their experience, sometimes for their unorthodox ideas, sometimes for their need to understand other

departments. The key is to put together in each team the right balance of strengths, experience, and points of view.

With the preparations complete, several pregame meetings are held to demystify the process and to encourage the teams to think about the substantive issues.

Zero Hour—The Simulation Begins

On the day the simulation begins the level of interest is high, and the competitive juices are flowing. After a short briefing to set the stage, teams go to team rooms to put their individual strategies into play. Teams maneuver aggressively for competitive advantage. Opportunities come and are seized, or are agonized over and are lost. Team activity intensifies. Deals are proposed, alliances fall apart, joint ventures are formed, intelligence is passed, rumors abound. What's important? What's in the noise? The game becomes real. One manager said, "I woke up last night in a sweat thinking what had happened in the game yesterday really happened."

The central control team of Booz·Allen simulation and industry specialists, along with the client and selected senior company managers, keep the play on course. Control arbitrates negotiations between teams, introduces external events, assists the teams, and resolves outcomes.

Drawing Up a List

As managers, we strive to refine our understanding of the competitive environment. We put the marketplace under a microscope and we record what we see. We take inventory of our resources. We shape a strategy. We draw up a list of contingencies that occur to us. We go back to the market and begin to apply our strategy.

But the elements that comprise the whole of the competitive environment are not static, nor do they exist on neatly parallel planes. They grow, they shrink, and they collapse. They converge, collide, rebound, and propel one another. They may appear from nowhere or drop out of sight. They are difficult to forecast. They can try our imagination. We cannot envision with any real accuracy how they might evolve even a few years from now.

A simulation is not a crystal ball, but for the first time, managers can develop and test ideas and alternative strategies whose human dimension makes them too complex to assess in any other way. At the same time it trains, it builds teams, and it unleashes the full intellectual power of the company on the issues.

A simulation takes all its participants simultaneously along one possible route into a future. Along that route they confront many of the situations they will confront on any route. Some will actually occur in real life. In those cases, the participants will have developed a strategy to deal with them.

As important, that experience has been shared by an influential cross-section of the company's managers. Going forward, they now have a common understanding of what may or may not work in a given situation, and they appreciate the elusive web of interdepartmental cooperation and support which will be key to any success.

Thomas Schelling, the preeminent American strategic philosopher and educator, in writing about the value of war games and simulations, said, "One thing a person cannot do, no matter how rigorous his analysis or heroic his imagination, is to draw up a list of things that would never occur to him."

In the end, that is the real value of a competitive strategic simulation. It generates ideas as yet unimagined.

Booz·Allen has worked with several oil companies to introduce the strategic simulation approach to resolve particularly difficult problems. The results have been very rewarding. However, the use of this approach, as with anything new, is very limited at this time.

THE IMPORTANCE OF CAPABILITIES

s discussed in Chapter 2, what determines how companies will perform has evolved over time. In the 1970s, the strategic concept was to pick attractive businesses. Attractiveness was defined as fast growing, profitable, etc. This was analogous to the rising-tide phenomenon. It's not that the company did anything right, it just got carried along with all the other lucky players in the vicinity. While appealing for its simplicity, it was not very practical. This concept was an influence in the formation of conglomerates. Conglomerates were simply bankers who funded attractive businesses. The fact that at any given time only 10%–15% of existing businesses are attractive did not seem to cause much concern.

As the conglomerate premise was defaulted, the strategic concept evolved to competitive advantage. Realizing the scarce number of attractive businesses, and their short life spans, companies began to realize that there was really no such thing as an attractive business, only attractive positions in businesses. Thus positional- based concepts, such as growth share matrices and Porter's framework, became the adopted standard for how to make money in a business.

Unfortunately, positional-based advantages are proving to be too transient. Many times, just as a company believes it has established an advantage, some competitor changes the rules of the game and the definition of advantage. The only way to react to this dynamic on a continuing basis is to build the capabilities within the organization to adjust. Capabilities are dynamic, positions are not. Capabilities are built along three business dimensions: time, cost, and/or value. For example, Honda and Toyota are considered to be premier auto companies. Yet they have built their capabilities around two different dimensions. Honda's strategy is to bring new innovations to market as quickly as possible. Conversely, Toyota's strategy is to have the lowest manufacturing and distribution cost. Both have been successful. They have succeeded not by having the positional best advantages but by embedding the organization with the capabilities to respond to externalities and still maintain their strategic course.

PERFORMANCE MEASUREMENT

ur experience has been that strategy development is not the primary driver of poor performance. The driver is usually strategy execution, which is becoming much more important to companies. One of the majors flaws observed in strategy execution is a lack of a performance measurement system for monitoring and course correction. Performance measurement is a key component of improving organizational effectiveness:

- Facilitates implementation of strategic direction
- Provides incentives for managers to act in best interest of corporation
- Establishes accountability
- Provides a vehicle to communicate what is important
- Provides measurable fair criteria to gauge the contributions of a business and individuals
- Provides feedbacks vs. targets in the goal setting, planning, and budgeting process

Recognizing these advantages, many companies are installing Performance Based Management Systems (PFMS). These systems are not elaborate systems or databases. They are based on a few critical performance measures for each business that provide reliable indicators of how and why each business is performing. Several tenets help guide a successful PFMS:

- Easily implementable and perceived as fair
- Provide focus for decisions
- Manageable in number
- Clearly linked to strategies and plans
- Forward thinking
- Measurable at a reasonable cost
- Linked to reward system

As part of performance measurement, value-based measures are becoming more popular. Essentially, value-based performance measurement is the scorecard from the shareholder's perspective. While shareholders are interested in the operating performance of the corporation, they are much more interested in how the overall yield of the stock (dividends plus price appreciation) compares with other investment alternatives on a risk adjusted basis. Shareholders always have the choice to sell the stock and invest in another corporation. Value-based measurement compares how a business has performed relative to what shareholders expect. Value-based planning takes this shareholder expectation one step further and states that corporations should only invest in businesses that have returns at or above that required by shareholders.

Measurement of value creation is straightforward for a public corporation. Its value is measured every day by its stock price. For individual businesses within a corporation, it is more difficult because they typically do not have observable prices. Value-based planning models have evolved enable measurement an individual business' contribution.

CONCLUSION

trategic planning should drive a company's business processes, organization structure and annual budgeting process, which in turn should establish resource levels and objectives for day-to-day operations. There are three approaches used by companies in developing strategic plans:

■ ***Bottom-up.*** Individual business units prepare plans, and the corporate staff simply aggregates them and subjects them to a critical reviews.

■ ***Top-down.*** In strongly centralized companies, a central planning unit prepares a single strategic plan with strong top management guidance, and lower echelons dovetail their own plans to the corporate plan.

■ ***Dynamic planning.*** Increasingly companies practice a more interactive approach that provides core guidance from the center but allows the plans to be developed in business units and then subjected to critical review by top company management.

As emphasized earlier, planning not only includes the development of optimal strategies, but increasingly planning also focuses on company's capabilities to implement these strategies. Some companies explicitly task their planners with organizational responsibilities, while others have created separate offices to lead these efforts.

Finally, the most important product of strategic planning is not a final plan, or budget, but rather the learning process which prepares managers to cope better with uncertainly, to deal better with the competition and their marketplace, and to respond more rapidly and appropriately to changes in the external environment. The process described previously does just that.

CHAPTER 10

OPTIMIZING THE VALUE CHAIN: MATERIALS AND MAINTENANCE CHALLENGES

There are two important aspects of petroleum company operations that traditionally receive insufficient senior management attention: materials management (procurement and materials handling) and maintenance. The two are linked here because there is a close relationship between the two activities: a significant percentage of purchasing and inventories are for the maintenance of existing facilities.

MATERIALS CHALLENGES

In a typical oil company, the purchase of goods and services from outside vendors (excluding crude oil feedstocks) exceeds 10% of total revenues and frequently offers a major opportunity for both cost reduction and the freeing up of valuable working capital (through reduced inventories). Taken together, oil and gas companies and their suppliers of services and materials constitute an "extended enterprise" that delivers finished product to customers (see Fig. 10–1).

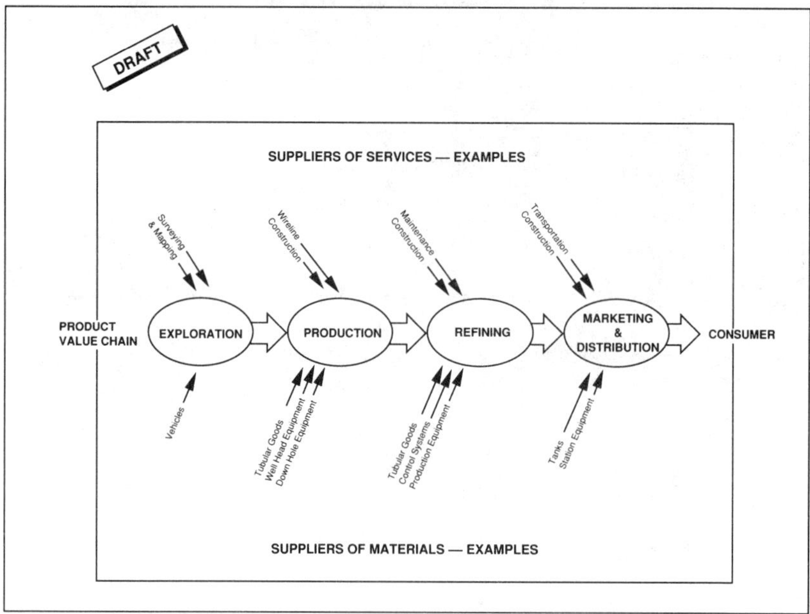

Fig. 10–1 Optimizing the Value Chain: The Extended Enterprise of Material and Service Suppliers.

In fact, much of the value added in this extended enterprise flows from the suppliers, not from the oil and gas companies themselves (see Fig. 10–2).

This section focuses on optimizing value by leveraging the capabilities of suppliers the part of the extended enterprise that exists beyond the organizational boundaries of the oil and gas company.

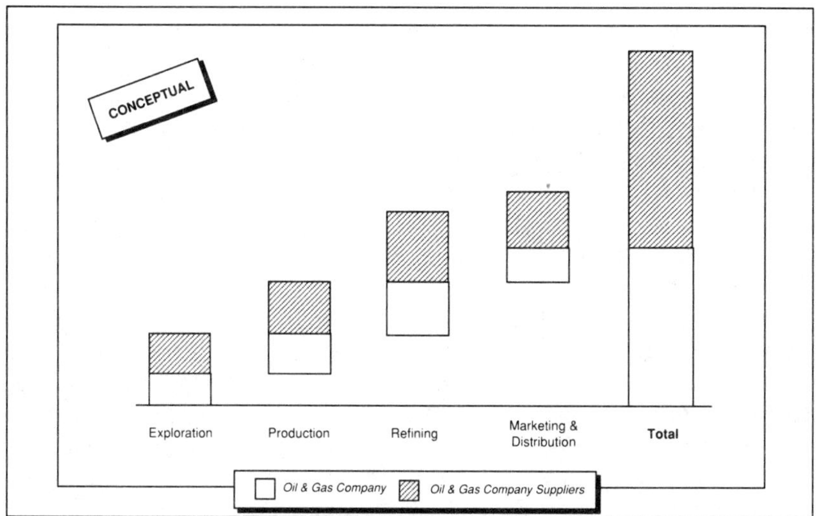

Fig. 10–2 Optimizing the Value Chain: Value-Added Structure of the Oil & Gas Industry.

Just like business process redesign, value enhancement is intended to ensure that non-value-creating activities are minimized. For example, many suppliers spend significant time and effort selling products and services to oil and gas companies. This effort (and cost) could be greatly reduced if an oil company decided which suppliers it would prefer to do business with and stopped making decisions one transaction at a time. For example, it is not unusual for drill bit salesmen to spend much of their time "selling" the local company man. As the drill bit supplier base has consolidated and oil companies have moved to longer-term contracts, this selling activity—which many would argue created little value—has diminished.

Resources must be mobilized to identify, prioritize, and unlock the value creation potential across the extended enterprise. Most oil companies still rely primarily on the purchasing department to keep the suppliers "in line." However, most purchasing departments are bogged down in cumbersome internal procurement processes and don't have the time to spend searching for new opportunities. Also, the purchasing professional may not possess the full range of skills needed. Successful companies explicitly plan and allocate cross-functional resources to manage their supplier's value-added activities.

Finding opportunities to better leverage suppliers' capabilities requires an understanding of the suppliers' position in the value chain. This position varies by the stage of oil or gas production. In exploration, the suppliers' value-added is primarily in knowledge-based services—surveying, mapping, etc. While in distribution/marketing, "extended enterprise" players range from ad agencies to builders of retail stores.

These suppliers add value in various ways. Value can be created through application of expertise in certain knowledge-based services—engineering design for example. Manufacturers/fabricators add value through leveraging their scale economies in their facilities and overhead. Distributors of equipment and supplies focus their warehousing and inventory management capabilities to spread costs and expertise over multiple customers.

Determining how best to optimize value with suppliers is therefore based on the influence an oil and gas company has on their suppliers. For example, pipe and valve distributors are direct suppliers to oil and gas companies, and many company purchasing departments have entered into "partnering" agreements to lower costs. However, a typical pipe, valve, and fittings distributor typically represents less than 20% of the cost of items bought, so value creation by lowering distributor costs is limited. The real issue is whether the oil and gas company has the ability to work directly with manufacturers to reduce the costs of the product supplied to the distributor (see Fig. 10–3).

Value chain optimization is hampered by the traditional relationships between oil companies and their suppliers. These relationships have long been at arms length, with the official single point of contact through purchasing. Unofficial contacts through engineering, maintenance, etc., have

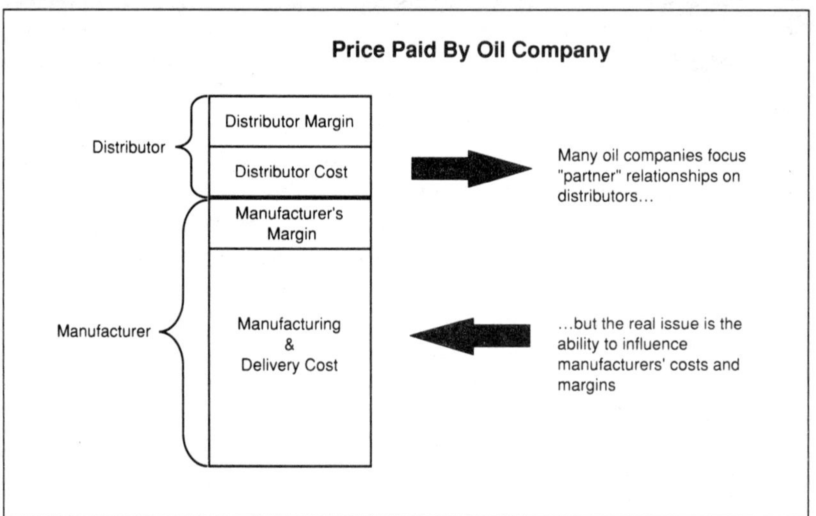

Fig. 10–3 Distributor Cost Structure: Price Paid by Oil Company

been discouraged by procurement policies. Purchasing's traditional objective—low unit prices—lead to transactional interfaces rather than a broader view of suppliers' capabilities.

Understanding the extended enterprise value chain is further complicated by the fact that one supplier may service the needs of many customers throughout an oil company. For example, a supplier of valves may be dealing with a project manager for new construction, the refinery engineering department for standards development , maintenance managers for field problem resolution, and materials managers for restocking inventories. This supplier fragmentation makes it difficult to identify and optimize their capabilities.

Oil companies and their suppliers often do not really understand each other. Suppliers are focused on the specific application of their offering and may not understand their relative value to the overall enterprise. Similarly, the fragmented customer is usually internally focused and spends no time trying to understand or take advantage of a potential suppliers' full capabilities.

In recent years oil companies have begun modifying their approach to suppliers, pushing for a tighter, more integrated relationship. This trend is driven by many factors, including:

- ***World-class procurement practices that were key to the success of Japanese car companies in building an extended enterprise.*** Consider Honda's relationship with its radio supplier, Alpine. Rather than specify a detailed design as U.S. car companies did for their suppliers, Honda provided Alpine with dimensions, desired functionality, and target cost. Alpine's engineers then designed the radio. That enabled Honda to focus its engineering attention on its piece of the value chain.
- ***Ever-increasing cost consciousness.*** As oil companies have gotten increasingly more cost conscious, they have begun looking more aggressively for cost reduction opportunities. Many oil company executives have only recently realized the magnitude of dollars spent with outside suppliers. They are turning to these suppliers for value creation opportunities. A second savings comes from business process redesign. As oil companies address their business processes, they are finding many of their capabilities may be redundant with their suppliers' capabilities (Honda's radio design department, for example). If those capabilities are not central to the business, they can be outsourced, producing substantial savings.
- ***Consolidation in the supplier base.*** As the oil industry retrenches, the number of suppliers of equipment and services shrinks. In drill bits, for example, there are now only a few surviving U.S. manufacturers. Consolidation has forced the survivors to become more creative and aggres-

sive in attracting and retaining long-term business. Suppliers are now more willing to work with customers in open relationships.

Many oil companies are adopting sourcing "best practices" from within and without the industry. These best practices include:

- A multifunctional approach that draws on the needs, experience, and unique expertise of individuals from every area of the organization affected by the sourcing process
- Understanding the extended enterprise—the entire supply chain from the materials suppliers to the ultimate end user
- Consideration of the total life cycle cost of goods and services, not just the purchase price
- Expanding beyond a traditional supply base to consider best practices and to move toward world class standards
- Commitment and continuing support from top-level management

A key element of applying these best practices is the closer relationships that have emerged in the industry between petroleum companies and suppliers – e.g., engineering & construction (E&C) firms. Over the last several years, several petroleum companies have formed closer alliances with E&C firms and OEMs. These alliances address a wide scope of activities, including;

- Engineering design
- Detailed engineering
- Procurement
- Construction
- Maintenance and modifications

Initial goals of these relationships focused heavily on cutting costs of major projects and shortening project schedules. The results of these closer relationships have also improved the engineering capabilities that are available during initial design as well as engineering fieldsupport. Companies have been able to better leverage the complete capabilities of the contractors to make their internal technical resources more readily available.

Results to date show that substantial benefits can be obtained in terms of quality improvement, cost reduction, and project schedule when these relationship exist. Companies and contractors alike report that the quality of the work and communication has increased significantly. Project schedules have become more dependable, and have been shortened by more than 5%. Safety on major project sites has also improved. Engineering quality, as gauged by errors, omissions and constructibility has improved, as has the level of innovation shown by the project teams.

Key aspects of these close supplier/customer relationships can be viewed this way:

GOAL:	Achieving specific business objectives by maximizing the effectiveness of each participant's resources
REQUIREMENTS:	Clear, mutually agreed on objectives to strive for world-class capability and competitiveness
EXPECTED BENEFITS:	Expected benefits include improved efficiency and cost effectiveness, increased opportunity for innovation, and the continuous improvement of quality products and services

These close supplier/customer relationships create opportunities to achieve substantial cost reductions by aligning the interests of owner and contractor. Direct project costs—engineering, procurement, construction, and rework—have been reported to be reduced by 5%–10% , with the savings split between owners and contractors. Various administrative costs have also been reported to reduced by up to 15% through streamlined communications processes. For example, legal costs are reduced on both sides by clearer, longer term relationships than by continued contracting and negotiating. Contractors are able to significantly reduce their costs of marketing and selling their services, as well as in preparation of project proposals. Costs associated with auditing and project accounting are diminished as jointly agreed upon time charging and invoicing procedures, which leverage available systems, are put into place and audit frequencies modified.

As discussed in Chapter 3, these relationships have moved companies from their traditional adversarial role with suppliers toward lifecycle part-

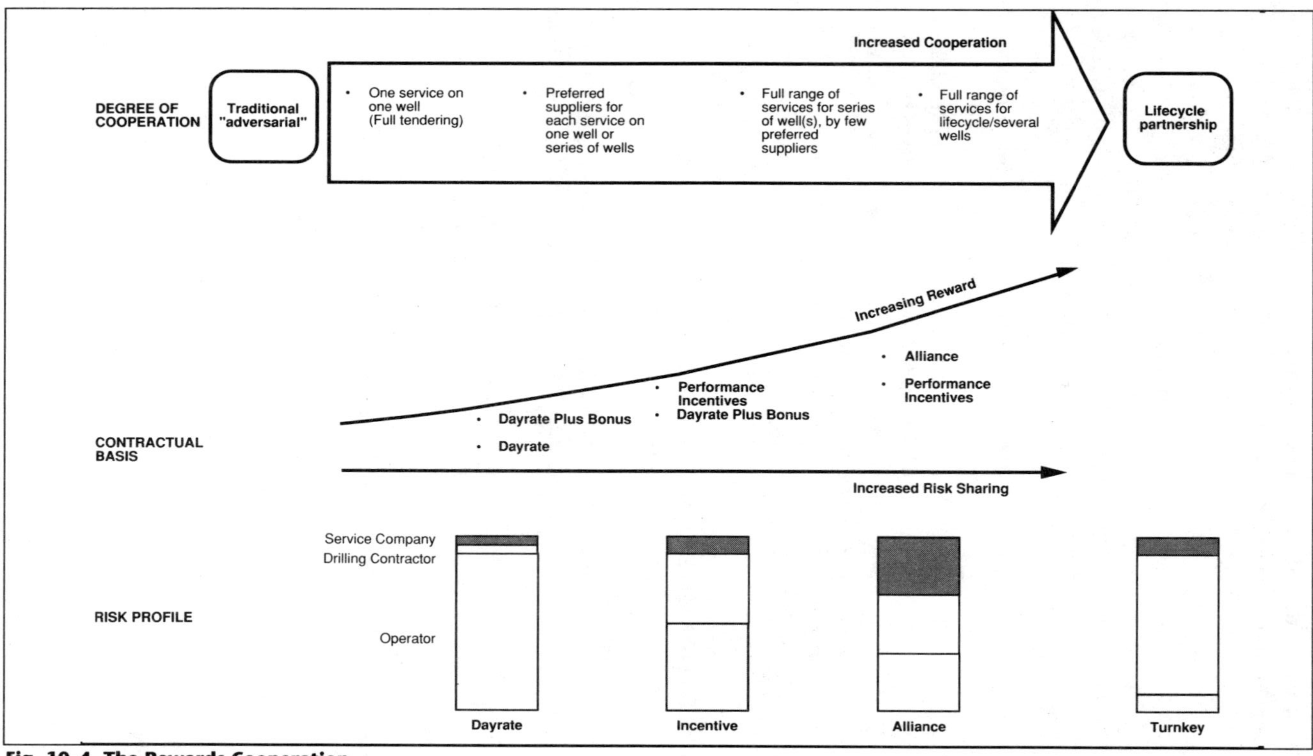

Fig. 10–4 The Rewards Cooperation.

nerships, with increasing rewards for superior performance by contractors (as well as commensurate increased risk) (see Fig. 10–4).

AN EXAMPLE

A recent oilfield development project undertaken by a global petroleum company is a good example of a new approach towards outsourced activities based on a contractual arrangement that allows contractor and client to share risks and rewards. This alignment was reinforced by creating a common commercial target and agreeing on a single goal for scope and time.

The scope of this project was for a gas wellhead platform and associated pipeline to the platform. The activities considered for contracting included all major project elements, from design through procurement, fabrication installation, and commissioning. The owner worked together with an alliance of contractors to complete the project.

The contractual basis was a short document covering:

- The development contract between the petroleum company and the alliance of E&C firms
- The alliance agreement among the E&C firms that laid down:
  - Scope of work
  - Conditions of satisfaction
  - Commercial basis
  - Risk/reward arrangement
  - Terms and conditions

All partners agreed on a target level of capital expenditure, exploration scope, and how overruns or underruns would be handled. The target capital expenditure was based on open book past performance with, a similar platform. Based on the ability to influence overall costs the risk/rewards were shared among the alliance members.

Using a composite team was another key to the successful integration of the various extended enterprise participants. A joint management organization was formed that consisted of a management team, various functional specialist representatives, and subproject leaders. The management team was made up of senior management representatives from each of the project partners. Functional specialists—e.g., Health, Safety, Environmental —reported directly to the management team. Subteam leaders—e.g., engineering—had the complete complement of resources needed to deliver their

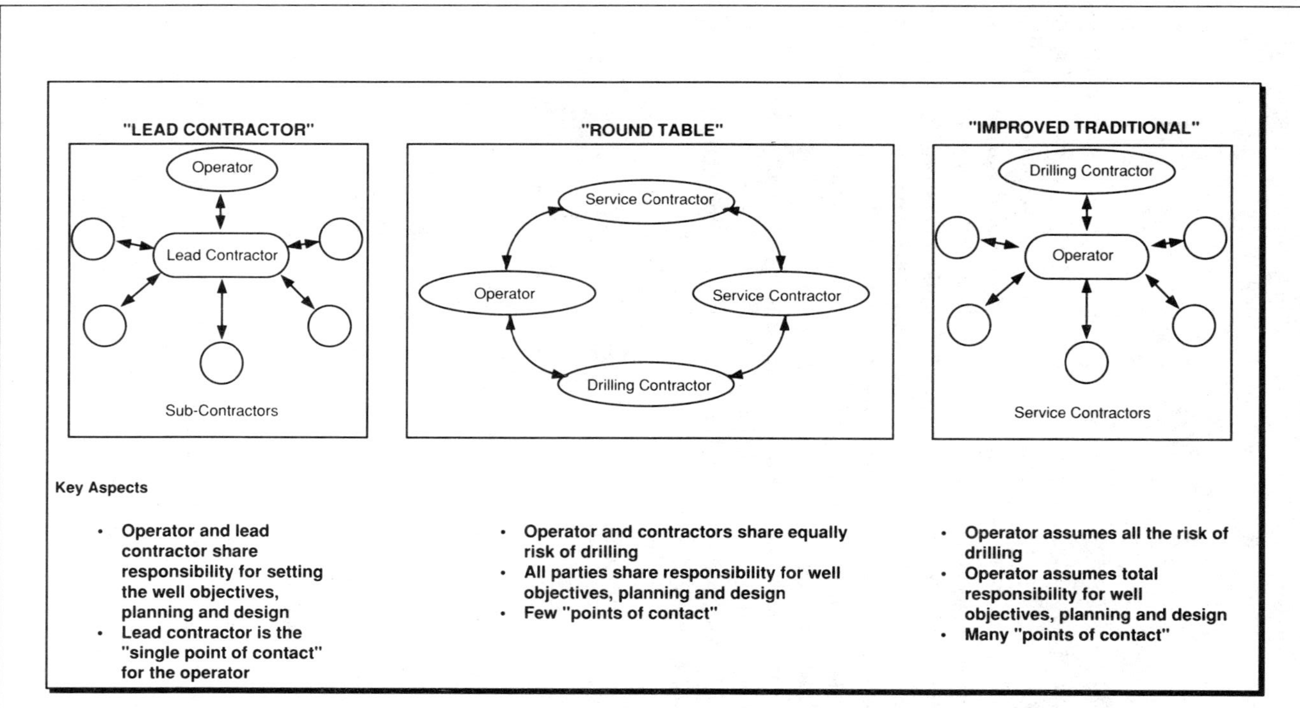

Fig. 10–5 Preferred Contractor/Operator Team Relationship

individual pieces. For each position the most capable person available was chosen from either the owner or contractor's organization.

This project organization reflected the "Round Table" approach discussed in Chapter 3 (see Fig. 10–5).

The atmosphere of trust and cooperation and the emphasis on common goals substantially reduced both cost and the time required to complete the project. That increased the profits for all parties involved. For example, because of close cooperation, substantial savings were attained by eliminating interfaces and duplication of effort. Additional upfront engineering was performed in order to reduce overall project expenditures. This also reduced the frequency and impact of design changes.

The results of this project highlight the potential for optimizing value across the extended enterprise. Costs were 25% below the agreed budget, with the millions in savings shared among the alliance partners (interestingly up-front engineering costs exceeded budget but resulted in lower costs—e.g., less rework—later on). The first commercial gas was delivered months ahead of schedule. This project schedule was significantly ahead of plan and much advantaged over recent, similar projects as well as the overall industry norm. Quality and safety measures were much better than expected for such a project.

This kind of approach can also be adopted to sourcing a single capability and can be used on small projects as well. That's exactly what one major petroleum company did when it wanted to reduce costs for the development of a small field. The company took several specific actions to get its contractors more involved. Representatives from the contractors (and operations) were involved early in the project (during the conceptual design phase) in order to get input about constructability and operations friendliness. Contractors were given functional specifications which allowed the company to take advantage of the contractors' know how. The company intends to use the same concept for a whole series of future small field developments.

A partnering relationship can also be built for the supply of a single capability. For example, one petroleum company determined that it did not have a strong capability for estimating costs during the conceptual design phase of a project. It decided not to build this capability in house. Instead, it developed a long-term partnering relationship with an E&C firm who is particularly strong in this area. On an as-needed basis, the company makes use of the E&C firm's capability.

The partnering concept can also be applied to modification and maintenance work, as illustrated by the outsourcing of these activities for another field. Both the operating and capital costs for this field increased substantially between 1980 and 1990. Operating costs increased fivefold, while annual capital costs rose by over 50%.

In order to cut costs, the owner decided to focus on core activities and outsource engineering support services to a limited number of contractors. An E&C firm was selected as the main service provider because of its proven ability to coordinate and execute integrated engineering and construction services. The owner was convinced that if traditional adversarial contractual relationships were improved substantially, cost savings would result. The E&C firm signed a partnering agreement with the owner. A single integrated team of client and contractor personnel was set up based on "the best man for the job" to provide engineering support to the field.

Initially, the team's involvement was limited to detailed design and procurement. Over time, the parties realized value could be created if they worked together more extensively. They modified the fee structure to reflect asset (e.g., equipment uptime) and team (e.g., budget)performance.

The partnering arrangement led to major improvements of business performance and substantial cost saving. The owner said that the quality of the engineering organization improved as well. Their responsiveness increased and the design turn around time was 30% faster than the traditional approach. The engineering process was simplified, and cost estimates improved as a result of better feedback from construction to design.

Another benefit of this approach is that it saved a lot of money– several million dollars annually. About half of these savings resulted from removing duplication of effort. Another 50% of the annual savings was achieved by improving productivity of offshore construction work, reducing late material deliveries, and working more closely with the offshore construction contractor.

The partnering relationship played a significant role in saving many millions and enhancing revenue as a result of wells being brought on more rapidly than expected.

A NORTH AMERICAN EXAMPLE

These techniques work for material and equipment suppliers as well. For example, one large North American producing company has adopted many of the emerging "best practices" in optimizing the value chain with its suppliers. Realizing that the process of acquiring materials and equipment for its drilling operations held enormous cost saving potential, the company launched a comprehensive study of all the processes involved in its materials acquisition and management functions. More than 100 interviews were conducted with representatives of key internal functions (e.g., engineering, drilling, maintenance, purchasing) at both headquarters and in field sites, as well as with a representative sample of the company's suppliers.

The study revealed much about the problems and consequences associated with the company's historic approach to managing across the extended enterprise:

- Constant reshuffling or churning of the supplier base, which limited efforts to develop sustained, mutually beneficial, supplier relationships
- A bid mentality for all expenditures that resulted in excessive use of purchase orders and small order quantities
- Inventories and safety stock far in excess of need, a result of both the field operations' lack of confidence in the existing purchasing function and an incomplete understanding of the true cost of downtime
- Breakdown in communication between field operations and the centralized materials management and purchasing functions at headquarters, creating a "we vs. they" mindset
- Incomplete understanding of supply economics and the concept of the extended enterprise
- The design and quality groups not knowing how their decisions affected material costs (e.g., engineers would try new piping, leading to new suppliers, excess stocks, and a new learning curve on remote site construction)

With the diagnostic results of the Materials Process Study in hand, the company created the following five recommendations to address the broad problems and the inherent inefficiencies:

- Create a new culture by involving senior management in setting policies and performance targets.
- Focus the organization on common goals by dismantling the geographic and functional chimneys that existed.
- Empower staff to execute the executive mandate by modifying tactical planning systems to increase rigor and discipline, both in procedures and systems.
- Drive results by launching formal teams to develop supply strategies and to capture purchasing, materials,

and inventory opportunities.

- Set the baseline for target stock levels and clean up excess materials.

To implement these recommendations, the company created a multidisciplinary core team composed of senior managers from the key organizational areas: the business units (field operations), central materials management, purchasing, and materials analysis. They were responsible for leading the materials process redesign efforts. They were charged with creating the structure, systems, and processes that would allow the company to implement the five recommendations.

To assist the company in realizing its strategic sourcing goal, the core team chose to apply the Sourcing Process shown in Figure 10–6.

In order to apply the model in a pilot program, the company's core team created two commodity teams and charged them with taking their assigned commodity families through the entire strategic sourcing process and leading the rollout of this new approach into their respective company units. "Commodity families" is the term the company uses to categorize the materials and equipment it purchases.

The commodity families chosen were valves (used in controlling the flow of oil and other fluids in the processing operations) and drilling commodities (primarily the piping and casing that go down the well, but also "downhole-jewelry " and wellhead fittings that control the well flow). These commodities were chosen for two primary reasons:

	TASKS
RESEARCH Develop and in-depth knowledge of the dynamics of the industry value chain	• Document current and future buys • Build a total acquisition cost model • Analyze industry structure and trends • Understand total cost drivers • Refine hypotheses for the next step, which is evaluation
EVALUATE Identify and assess various sourcing strategies and evaluate selected suppliers' capabilities to deliver against them	• Examine process and product technology tradeoffs • Develop the extended enterprise model • Screen selected suppliers • Identify and quantify key business process opportunities • Create implementation action plans
STRUCTURE Develop supply relationships and action plans to build the relationship infrastructure	• Develop target cost models • Prepare business proposition and present to target suppliers • Select sourcing partners (suppliers) and define commercial agreements • Develop supplier action plans • Prepare final action plans for developing infrastructure

Fig. 10–6 Strategic Sourcing Process for Optimizing Value Across the Extended Enterprise.

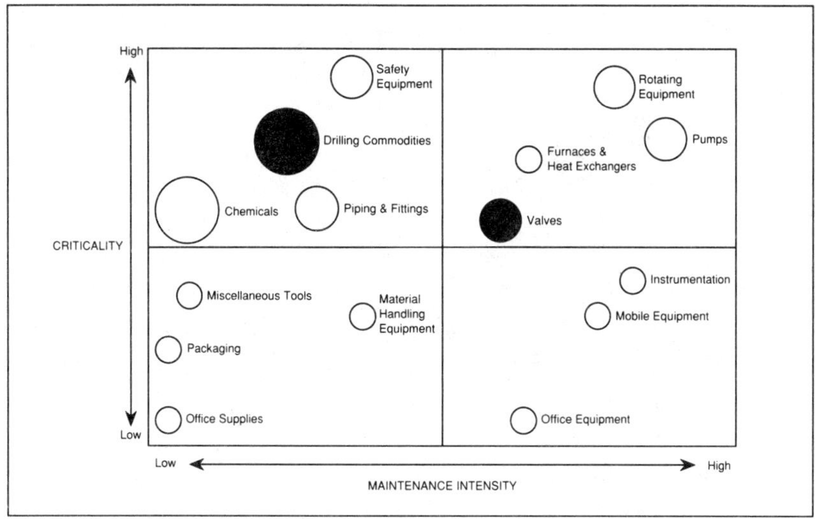

Fig. 10–7 Commodity Segmentation Model.

- These ranked high in terms of their importance to sustaining operations and represented a significant portion of the total materials budget (see Fig. 10–7).
- Distinct differences existed between the families in terms of purchasing and materials management. That would challenge and test the applicability of the full strategic sourcing process across a range of commodities.

The teams created to carry out the strategic sourcing pilot program were composed of members who represented the complete range of functions involved in the full sourcing process, from initial design to the acquisition of commodities and their installation, use, and maintenance. For example, the drilling commodities team included individuals from drilling operations, drilling engineering, purchasing, and materials management. The valve commodities team had similar representation.

These teams began by evaluating the company 's historic approaches. For example, the Drilling Commodities Team research showed that over the last five years , the company had bought from 17 different suppliers and that annual purchases from each fluctuated significantly from year-to-year. They concluded that this churning of the supply base greatly inhibited the company's ability to sustain improved relationships with suppliers, and that it also resulted in excessive inventories and left suppliers unable to forecast demand. The result was that often they were unable or unwilling to meet it.

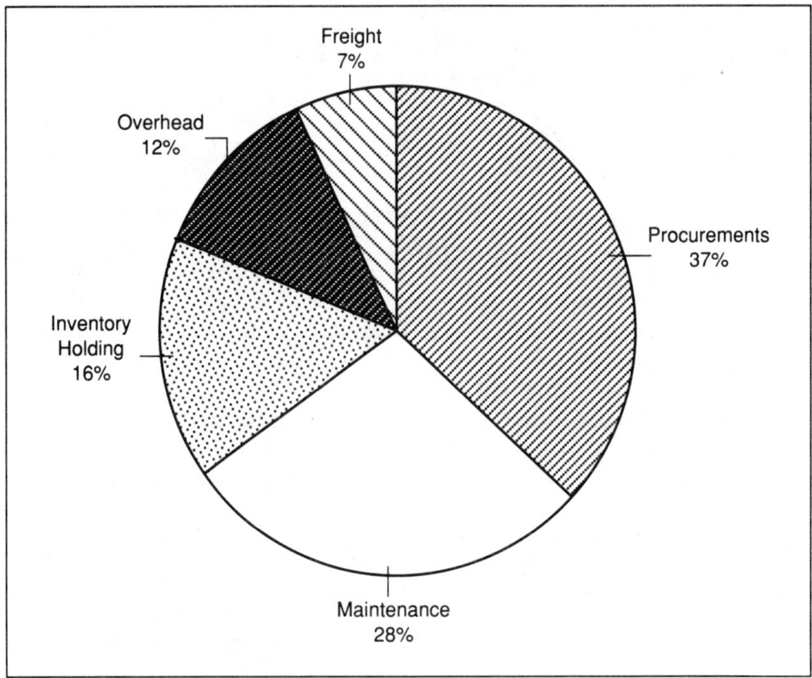

Fig. 10–8 Total Life Cycle Cost Model.

The teams built an understanding of the total life cycle costs, quantifying all the costs of materials acquisition (often some that had not been recorded). These go beyond the invoice price and can include the costs of quality rejections, carrying inventory, transportation, material acquisition and handling costs, and internal downtime when material is unavailable. The total cost acquisition model developed by the Valve Team is shown in Figure 10–8. In this case, the initial purchase price represented only 37% of the total life cycle costs of valves.

One of the Drilling Commodities Team's particular concerns was whether it was more advantageous to buy directly from the manufacturer instead of through distributors. They wanted to determine which types of suppliers would be the right fit, given the company's purchase volume and supply requirements. The industry structure model they developed (see Fig. 10–9) helped them eventually to decide on siding with distributors, where they determined they could create the strongest relationships and achieve better long-term pricing, as much as 5% lower.

Understanding total cost drivers was also a key element. For example, in the company's case, the Drilling Commodities Team hypothesized,

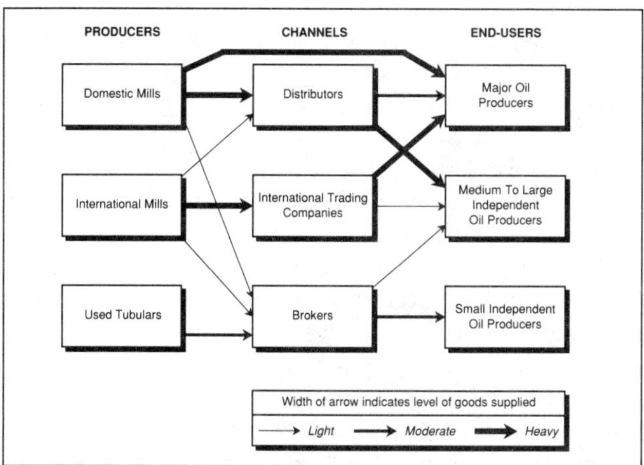

Fig. 10–9 Tubular Goods Industry Model.

freight costs for tubulars could be lowered if the need for expediting shipments could be reduced. They identified the factors that contributed to freight premiums paid in the preceding year (see Fig. 10–10) and analyzed the reasons for expediting shipments. They projected that 3%–5% of total freight costs could be saved by better materials requirement planning and stronger distributor partnerships.

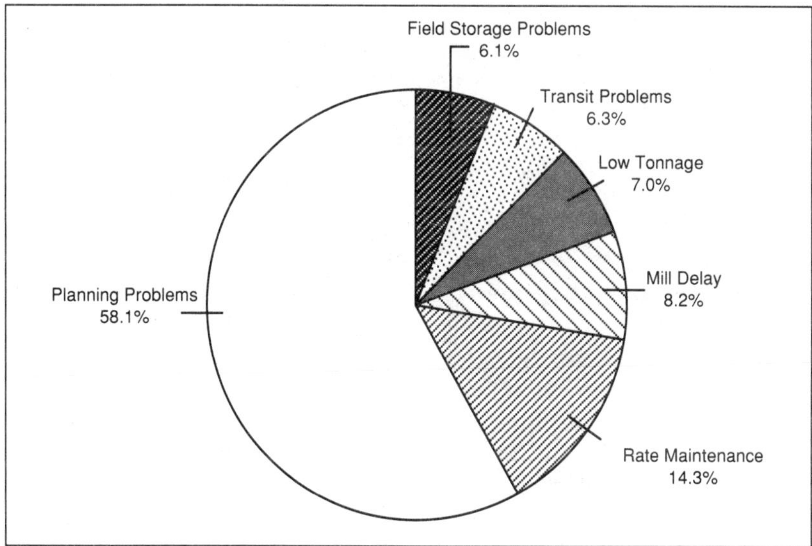

Fig. 10–10 Reasons for Paying Freight Premiums: Tubular Goods.

Revising specifications was another key lever that the company addressed to better optimize the extended enterprise value chain. When the company began drilling in the Arctic, there was very little experience in operating in such a hostile environment. As a result, materials specifications were developed very conservatively. Experience, however, demonstrated that some of the initial specifications were excessive. The Drilling Commodities Team examined such a specification for certain casings for extended-reach wells. The team was multidisciplinary, including both drilling operations and engineering representatives, and so was able to determine that one of the current casing specifications was overly conservative and that a lighter grade would be sufficient. Switching to the lighter grade saved 2% on the casing costs (see Fig. 10–11).

The teams also identified opportunities by finding ways of better integrating with and leveraging supplier capabilities. For example, the Valve Team began by analyzing sourcing practices for the three types of valves the company purchased. They found that purchases were being made from quite a large number of suppliers and that minimal effort was expended in building value-added relationships with them. From this analysis, the team defined an extended enterprise model that included the supply base, and identified its value-added potential (see Fig. 10–12).

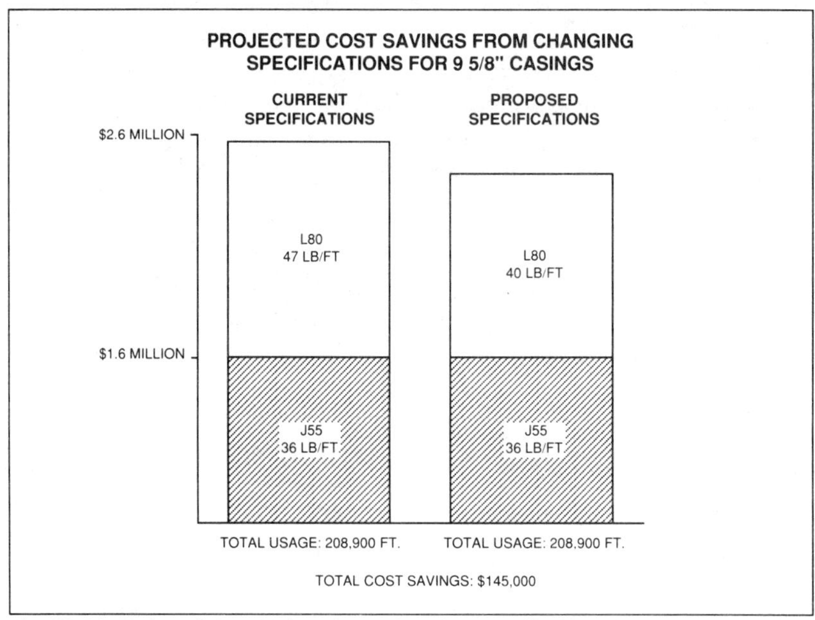

Fig. 10–11 Projected Cost Savings from Changing Specifications for 9 5/8-in. Casing.

Extended Enterprise Element	TYPE OF VALVE		
	Ball/Plug	**Choke**	**Forged Steel**
Supply Base	• Reduce from twenty suppliers to three or four • Use local distributor to provide repair parts for plug • Develop local repair shop	• Reduce from four manufacturers to one or two • Develop new source for repair parts by moving upstream in supply chain	• Reduce from four distributors to one • Leverage available only at distributor level
Relationship Structure	• "Extended relationship" with target suppliers • "Dedicated supplier" relationship with repair shop	• Work toward true comaker relationship with core supplier • Apply "lifetime agreement" with formula-based pricing long term	• Try to develop "full-service distributor" relationship • Quote annual or multiyear quantities and direct business accordingly
Value-Added Services	• Advanced metallurgical support • Hold all hard goods inventory • Maintain documentation files for ready access	• Continued engineering support of new designs • Stock all hard goods • Maintain documentation files for ready access	• Improved coverage of stocked items • Convert threaded ends to sockets as required
Supplier Integration	• Facility engineering to drive specification changes • Supplier engineering with slope maintenance for preventative maintenance • Purchasing with manufacturing engineering to drive down electroless nickel plating cost • Contractors and suppliers to control specifying	• Corrosion group to work with supplier designers to improve durability • Material specialists to drive planning • Quality assurance to eliminate source inspection • Contractors and suppliers to control specifying	• Material specialists to drive planning • Facility engineering to drive specification changes Contractors and suppliers to control specifying

Fig. 10–12 Extended Enterprise Model: Valve Commodities Team.

The results continue. Examples include:

- The Valve and Drilling Commodities Teams have documented at least a $5 million cost savings in the first 18 months of implementation.
- In the roll-out of the second wave of applying the strategic sourcing cycle, the Chemical Commodities Team has projected an estimated $7 million in cost savings.
- Excessive inventories of drilling supplies have been reduced from $12 million to $2 million.
- Many supplies and parts have been standardized, further contributing to efficiency.

Optimizing the value across the extended enterprise requires new approaches in goal setting, corporate culture, supplier interface and capabilities (see Fig. 10–13).

As with any major change, this is easier said than done. However, oil and gas companies who are implementing these new approaches have concluded that the value optimization benefits far out weigh the potential implementation barriers.

MAINTENANCE CHALLENGES

The acquisition cost leverage of the feasibility and design stages encourages attention to equipment and process design, but minimizing life cycle costs also requires attention to equipment maintenance . Life cycle cost, even more than acquisition cost, is beginning to drive contracting decisions for long-term investments in process plants. Maintenance-related design phase activities include design for easy maintenance and common equipment. After commissioning the maintenance focus shifts to differentiated asset management.

Design for maintainability requires participation from engineering, maintenance, and operations. These functions can jointly optimize the main design factors influencing maintainability: intrinsic reliability, physical accessibility, new tool/technology requirements, and part count. For example, mechanical seals have higher acquisition costs than traditional packing, but they reduce shaft wear, energy usage, water consumption, and maintenance labor requirements. Physical accessibility should be considered for both repair/changeout and condition monitoring. In addition, a

TRADITIONAL PRACTICES	NEW APPROACH
Goals and Objectives • **Each party's goals** and **objectives,** while similar, geared to what is best for them	**Goals and Objectives** • **Shared goals** and objectives ensure common direction
Culture • **Suspicion and distrust;** each party wary of the motives of actions by the other • **Objectivity limited** due to fear of reprisal and lack of continuous improvement opportunity • Normally **limited to project level** personnel	**Culture** • **Mutual trust** as the basis for strong working relationships • **Objective critique** geared to candid assessment of performance • **Total company involvement** commitment from CEO to team members
Interfaces • **Communications structured** and **guarded** **Limited access** with structured procedures and self-preservation • taking priority over total optimization **Sharing limited** by lack of trust and different objectives • **Duplication** and/or translation with attendant costs and delays	**Interfaces** • **Open communication** to avoid misdirection and bolster effective working relationships • **Access** to each other's organization; sharing of resources • **Sharing of business plans** and strategies • **Integration** of administrative systems and equipment
Contracts • **Not innovative** strategies – Large number of contracts, comfortable with current arrangements – Manpower intensive, large number of interfaces – Constrained by interpretation of legislation Single project contracting • Single project contracting • Routine **adversarial relationships** for self-protection • For, **cumbersome** in-house **processes** e.g. procurement, contracting, etc......	**Contracts** • **Innovative strategies** – **Fewer contracts** – **Fewer interfaces** – **Less manpower** **Long-Term commitment** provides the opportunity to attain continuous improvement • Absence or minimization of **contract terms** that create an • adversarial environment • **Streamlined processes** e.g., procurement by the contractor
Capabilities/Project Management Team • All disciplines represented in our project team **"checking role"** • Separate project teams **"policing the contractor"** • Application of **standards**	**Capabilities/Project Management Team** • Focus on front end capabilities and high level guidance and control **small** integrated project **teams** • Use of functional specification and industry standards

Fig. 10–13 New Approaches to Strategic Sourcing.

layout minimizing personal protection requirements (e.g., respirator vs. full suit) improves the capabilities and efficiency of personnel making repairs or performing preventive maintenance.

Individual plants often ensure a maintenance role in capital project review, but many greenfield, corporate, or contractor projects have no maintenance review mechanism. Incorporating maintenance personnel into the design: build teams or holding design reviews for maintenance personnel addresses this deficiency. Experiences from pulp and paper to aerospace support the cost effectiveness of maintenance review activities.

Part count and new tool/technology requirements affect life cycle cost and focus attention on equipment commonality. Commonality reduces maintenance labor costs by narrowing the training required for mechanics and operators, allowing them to move more quickly down the learning curve. Commonality also dramatically reduces maintenance inventory stocking levels, particularly in critical spares which often sit unused for years. It also allows purchasing to negotiate purchasing price reductions, and the increased volume creates opportunities for supplier stocking and consignment. Finally, commonality reduces artificial obsolescence of spare parts by minimizing churning of new equipment through the plant.

Operating consistency and performance is enhanced by common equipment. Improvements are often elusive when organizations attempt to fit special equipment and/or processes into their normal operation procedures. For example, Kodak found that replacing 20% of pneumatic control systems in a facility with state-of-the-art programmable electronic controllers had no impact on either batch yield or yield variability. Exploiting electronic controller capabilities required new operations and maintenance processes which the organization was not prepared to adopt as long as pneumatic controls were the dominant system.

Achieving equipment commonality in the face of evolving technology (often with fundamentally superior performance or reliability) is a formidable task. Central equipment control stifles continuous improvement, yet lack of central control leads to a mishmash of equipment both within and across plants. Intel's "copy exactly" program is an example of accomplishing both central control and continuous improvement. Design, installation, and upgrade authority is centralized and given the power to both forbid and demand changes to equipment and processes. A decentralized continuous improvement mechanism supports the central authority, however, focusing on (1) generation of new ideas and proof of concept at the plant level and (2) rapid dissemination of proven ideas to all plants.

Design for maintainability and equipment commonality ensures that ongoing maintenance activities are not structurally disadvantaged in mini-

TYPE	$/HP/YR	DESCRIPTION
Breakdown	$17.50	• Run machines to failure
Preventive	$12.00	• Calendar base (MTBF) replacements – Doesn't catch all of the breakdowns – Replacing some parts after they have been causing product quality, availability, or throughput reductions – Replacing some parts not in need of replacement – Introducing re-build/re-assembly errors
Predictive	$8.00	• Replace only parts needing replacement – Based on trending machine parameters such as vibration – Does not include root cause analysis of failures
Precision	$4.50	• Replace only parts needing replacement • Machinery improvement program to eliminate root causes of equipment failure, e.g., forced vibration reduces bearing life; addressing causes of vibration allows bearing reach expected life

Fig. 10–14 Budget for Various Maintenance Philosophies.

mizing life cycle costs. Ongoing maintenance activities ensure that the life cycle cost potential is met, and often make up 15%–25% of operating budgets, with these dollars split evenly between labor and materials. Maintenance activities are also a prime driver of the equipment effectiveness (incorporating availability, throughput rate, and yield) of existing assets.

Optimizing operating maintenance lowers maintenance expenses and increases equipment capacity. To a large extent, maintenance philosophy (all else being equal) determines the maintenance budget (see Fig. 10–14).

The shift from corrective to preventive/predictive maintenance activities drives reductions in maintenance expense and increases equipment capacity. Unit cost reductions begin immediately, while maintenance budgets begin to decline after roughly two years. Industry typically expends 60% of maintenance labor on corrective activities, and only 40% on preventive/predictive tasks. Consequently, most maintenance organizations excel at addressing sporadic problems resulting from changes in equipment condition: they simply restore the equipment to its previous condition. Chronic losses are often hidden or perceived as normal, though they may offer more potential improvements in equipment effectiveness than sporadic losses. Addressing chronic losses requires detailed root cause analysis, often involving a fundamentally different look at sources. For example, a petrochemical team looked at pump failures—involving seals, shafts, impellers, housings, and couplings—and concluded that the failures were *all* due to "special" or "one-of-a-kind" causes. This team was a captive of the existing database, which categorized failures by equipment type (pumps in this case) and failing element. A second team looking at the same failures determined that *all* the failures could actually be traced back to either alignment or seal installation.

Maintenance philosophy and activities must also recognize the need for differential equipment maintenance strategies, i.e., different pieces of equipment have different maintenance needs. Differentiation should be based on actual root causes of equipment degradation. For example, equipment suffering from vacuum leaks needs ultrasonic testing rather than vibration analysis, and a 50,000-rpm turbine requires more frequent monitoring than a 250-rpm paper machine roll.

Equipment criticality and the cost of downtime also play a role in prioritizing the frequency and intensity of preventive/predictive activities. Many companies implicitly recognize the importance of large, critical equipment and concentrate their preventive/predictive and root cause analysis there, without realizing the impact of smaller equipment failures on machine performance and maintenance budget (see Fig. 10–15).

For example, at one company, an electrician spent half of every day "laying hands on" all 2400V motors in the plant. At the same time, the leading cause of failure for motors below 100 HP was lubrication: motors were never regreased after installation. Organizing maintenance activities to include smaller equipment is a key element of best practice maintenance.

Recognizing the importance of predictive and preventive activities deemphasizes the scale benefits of central maintenance organizations in favor of the responsiveness, equipment familiarity, and operator/mechanic relationships of area maintenance organizations. The next logical step inte-

EQUIPMENT TYPE	PERCENT OF PLANT MACHINES	MAINTENANCE $/HP/YR	COMMENTS
<500 HP	5-10%	$11	• Usually running precision already • Typically process critical equipment without spares
100-500HP	~30%	$22	• Usually running "OK" • Lower maintenance effort focused on preventive or predictive philosophy
<100 HP	~60%	$49	• Often employ "breakdown" maintenance • Typically most of the equipment in repair shops Repair/rebuild cost similar to larger equipment – Smaller machines have smaller absolute tolerances – Parts cost a similar amount regardless of size – Amoco found that pump rebuilds cost $4,000-$6,000 per pump regardless of size Often 80% or more of maintenance budget spent on low HP equipment

Fig. 10–15 Impact of Machine Failures on Maintenance Budget.

grates maintenance and operations activities. Tennessee Eastman found that 40% of the traditional mechanic's work could be performed by an operator with minimal training. An operator could perform another 40% with additional training. Only 20% of maintenance tasks actually required a certified mechanic's skills. Operators were not converted into maintenance personnel, but they perform basic, routine preventative maintenance tasks such as cleaning, changing filters, bolting, lubricating, adjusting, etc. Currently roughly 80% of over 3,000 maintenance related tasks are now being performed by operators. Tennessee Eastman found that 75% of maintenance problems could be prevented by operators at an early stage and that integrating maintenance activities into manufacturing increased equipment availability by 1%–2%.

A petrochemical company found that field operators' regular duties required only 50% of their time, with half of the remaining balance available for performing preventive maintenance tasks. Another client eliminated 75% of the maintenance organization by replacing operators with mechanics (who moved from maintenance to operations). The mechanics are better positioned to prevent and troubleshoot failure on machines they operate, and are also responsible for teaching their new colleagues preventive maintenance techniques. The core maintenance group is now limited to high skill activities (e.g., vibration analysis, machine shop, electronics) and activities concurrent with line operation (e.g., rebuilds, buildings and grounds).

CONCLUSION

Taking an integrated view of the value chain from procurement through maintenance is an essential strategic capability. The result will be lower costs, shortened cycle time, improved quality and reduced working capital (by reducing inventory). This approach will draw on the techniques of business process redesign discussed earlier and is also linked to the concept of the extended enterprise, discussed later.

CHAPTER 11

QUALITY MANAGEMENT

While major improvements in performance are best accomplished by adapting new strategies and redesigning business processes, as discussed in previous chapters, the process of continuous improvement and of ensuring consistent quality in all aspects of corporate endeavor is an ongoing, daily management challenge. Grasping either the importance or relevance of quality programs, quality management, quality systems, the quality movement, or total quality management (TQM), is understandably difficult. Given the prevalent and convenient use of jargon found in the literature of business and academia addressing the subject of quality, it is little wonder why there exists such ubiquitous confusion and skepticism towards quality management. Moreover, quality management is not a homogenous set of concepts. Quality management, as defined by formal standards such as ISO 9000, is not the same as TQM. Even more problematic is the fact that TQM does not derive from one common standard, with the exception of the Malcolm Baldrige National Quality Award (MBNQA) criteria, which makes it difficult to analyze (see the appendix for additional background material).

Evidence indicates, however, that formal approaches to quality management of varying description and prescription, are widespread and growing. A 1992 study by Ernst and Young and the American Quality Foundation of 584 large American, Canadian, German, and Japanese companies in four industries found that more than half of these companies' employees participate in meetings about quality. Additionally, standards for quality management continue to be updated and created, such as the oil industry's API Q1 and the international standard ISO 9000 Series. Quality awards, the criteria of which often serves as a *de facto* quality management standard, continue to increase in number and industry scope. The demand for the Malcolm Baldrige National Quality Award (MBNQA) guidelines grew rapidly in the late 1980s and early 1990s, although the quantity of both award applicants and the quantity of guidelines requested appears to be going down (see Fig. 11–1).

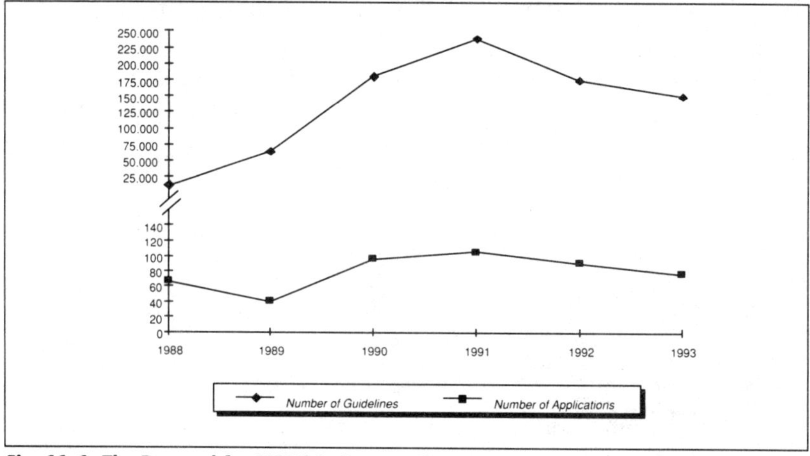

Fig. 11–1 The Demand for MBNQA. Source: The American Society for Quality Control (ASQC).

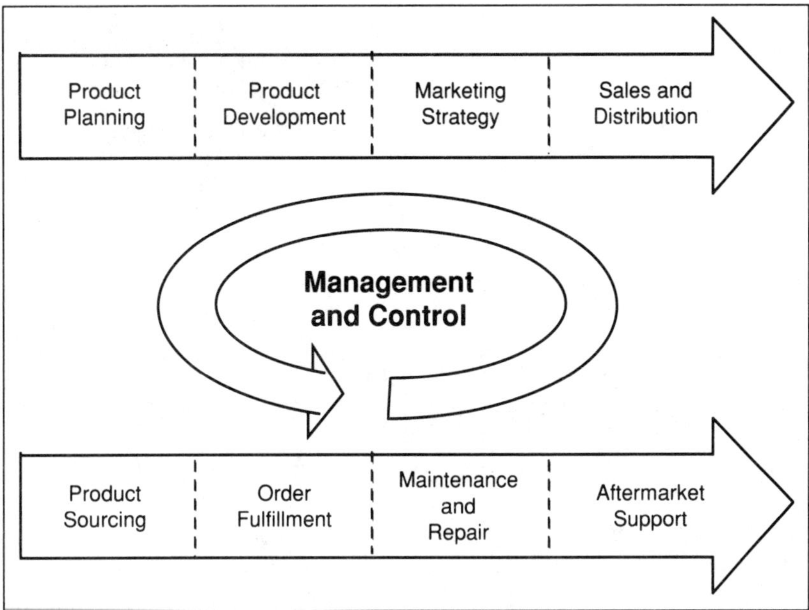

Fig. 11-2 Management and Control Through Quality Management.

Leading business journals, such as the Harvard Business Review, Sloan Management Review, and the California Management Review, are devoting more writings to the subject of quality management, mostly in terms of TQM, i.e., customer satisfaction, continuous improvement, and employee empowerment. Additionally, academicians from the nation's leading business schools, such as David Garvin, Michael Jensen, and Karen Wruck of the Harvard Business School and George Easton of the University of Chicago Business School, are helping to clarify and to advance the subject matter of quality management.

While some advocates claim that quality is free, in our experience, not all quality programs are free: many have simply driven up costs and reduced effectiveness when an overly bureaucratic "quality program" is imposed on a skeptical organization.

In this chapter we seek to establish some substance and coherence to the role of quality management within profit-seeking organizations, with particular insights pertaining to the idiosyncratic quality management practices of the oil industry. As discussed in greater detail in this chapter, the results associated with quality management implementation within the oil industry and across all industries are quite mixed. Rather than speak *ad nauseam* to the technical details of quality management standards, however, we devote some discussion to the strategic importance of some quality management concepts, integrating and building on leading-edge research in economics, finance, and organizational theory, as well as empirical evidence from industry. These bodies of knowledge provide useful constructs for understanding how certain quality management concepts represent a contemporary technology for organizing the company, addressing the fundamental decision making and control issues affecting the company, helping to provide management and to control over the company's internal capabilities (see Fig. 11–2).

FUNDAMENTAL ORGANIZATIONAL PROBLEMS AFFECTING THE COMPANY

The challenge for management is how to balance the benefits of decentralization of information and decision making while ensuring that shareholders' interests are served. Throughout the history of quality management, from the early days of inspection to TQM, quality management served as the mechanism by which specific knowledge pertaining to product, service, or process integrity was created and used. Quality management also served as a system of internal controls necessary to mitigate the self-interested behaviors of workers perversely incentivized by criteria other than "quality." These internal control mechanisms provided by quality programs usually take the form of control processes (see Fig. 11–3) or oversight authorities (e.g., inspectors and auditors). What has changed throughout the evolution of quality has been the quality management methods used to accomplish both the dual roles of knowledge creation and use and internal control.

There are four distinct and evolving forms of quality management: quality management as inspection, statistical quality control, quality assurance, and strategic quality management. Quality management in the United States has expanded in ever-widening circles, each era incorporating elements of the one that preceded it.

As simple as its name implies, *inspection,* originating in the eighteenth and nineteenth centuries, occurred as the post-production method of assuring that products met the customers requirements. Inspection methods advanced from the matching of parts with one another by hand during low-volume manufacturing, to more formal methods making use of jigs, fixtures, and gauges. The industrial revolution, with mass production and the need for

GENERIC ELEMENTS	NQA - 1 18 Elements	ISO-9001 20 Elements
Management Responsibility	Organization	Management Responsibility
Design Control	Design Control	Design Control
Quality Program	Quality Assurance Program	Quality System Training
Services	Procurement Document Control	Purchasing Servicing Contract Review
Material Traceability	Identification & Control of Items Control of Purchased Items & Services	Product Identification and Traceability Purchase Supplied Product
Documentation	Document Control Instructions, Procedures, and Drawings	Document Control
Special Processes	Control of Special Processes	Process Control
Inspection	Inspection Inspection, Test, & Operating Status Test Control	Inspection & Testing Inspection & Test Status
M&TE	Control of Measuring and Test Equip.	Inspection, Measuring and Test Equip.
Handling & Storage	Handling & Storage	Handling & Storage
NCR Control	NCR Control	NCR Control
Corrective Action	Corrective Action	Corrective Action Statistical Techniques
Records Control	Quality Records	Quality Records
Audits	Quality Audits & Surveillances	Quality Audits & Surveillances

Fig. 11-3 "Control by Process" Standards.

interchangeable parts, drove the need for advances in inspection methods. Frederick Taylor gave the inspection activity added legitimacy in the early 1900s by singling it out as an assigned task for one of the eight functional bosses (foremen) required for effective shop management. Inspection activities were linked more formally to quality control in 1922, with the publication of G. S. Radford's *The Control of Quality in Manufacturing*. Quality was viewed, for the first time, as a distinct management responsibility and as an independent function. Here practices stood for several years.

Statistical quality control originated in the next decade, stimulated by research conducted at Bell Telephone Laboratories. In 1931, W.A. Shewhart's *Economic Control of Quality of Manufactured Product* was published, giving the discipline a scientific footing for the first time. In his work, Shewhart gave a precise and measurable definition of manufacturing control, developed powerful techniques for monitoring and evaluating day-to-day production, and suggested a variety of ways of improving quality. The application of the principles of probability and statistics to enable process control and sampling were the genesis of statistical quality control. Surprisingly, however, neither sampling techniques nor process control charts had much of an immediate impact outside the Bell System until World War II. Little would change until several key works were published in the 1950s and early 1960s. These ushered in the next major quality era, that of quality assurance.

Quality assurance represented an evolution of quality management from a narrow, manufacturing-based discipline to one with broader implications for management. Problem prevention remained the primary goal, but the tools expanded far beyond statistics. Joseph Juran's first edition of his *Quality Control Handbook* in 1951 began to address the economics or costs of quality, tackling the critical question senior managers faced, i.e., how much quality is enough? The cost of quality measure brought forth by Juran generated much information managers needed to understand costs and productivity and highlighted the importance of designing the right product right the first time.

Armand Feigenbaum's first edition of his *Total Quality Control* in 1956 took the cost of quality principle a step further by proposing that quality products were unlikely to be produced if the manufacturing department was forced to work in isolation. The underlying principle of Feigenbaum's total quality view was that to provide genuine effectiveness and efficiency, control must start with the design of the product and end only when the product has been placed in the hands of the customer who remains satisfied. The first principle to recognize is that quality is everybody's job. Feigenbaum noted that all new products, as they moved from design to market, involved roughly the same activities. From a quality management standpoint, they could be grouped into three categories: new design control, incoming material control, and product or shop floor control. These control activities are the hallmark of so many quality standards that rely heavily upon the "control by process" and audit/independent oversight approach to ensuring conformance to requirements.

Strategic quality management is where formal quality management approaches tend to reside today. In other words, formal quality management approaches now are written to provide strategic importance to the company, in terms of providing management and control over its core capabilities, focused on either conformance to requirements or customer satisfaction, or both. Much of what is called quality management today is essentially representative of quality assurance and quality control (QA/QC) practices devised in the 1950s. In fact, it wasn't until recent years (1987 to be exact) that several of the related QA/QC standards were written into one common standard, the ISO 9000 Series.

Admittedly, there have been some modifications to the standards, but the underlying emphasis on control and the philosophy for control has remained relatively steadfast. Internal controls, as listed in Figure 11–3, are reflected in terms of process "checks-and-balances" (e.g., design control), documented work procedures, certifications, and audits to the procedures, as well as the independent oversight of a separate organization (usually a formal quality function). The concept of TQM, however, is a distinctly different concept addressing quality management, particularly in terms of the philosophy of

	QA/QC	TQM
Knowledge Creation and Use	Corrective Action methodologies required by ISO 9001 and other equivalent standards which typically suggest some form of root-cause analysis. Participants usually limited to those involved with the discrepancy identified.	Use of the scientific method for problem solving; emphasis on placing the scientific method skill set in all employees. Steering Comittees require teams to provide clear evidence of applying the scientific method to their problem solving efforts.
Decision Right Assigment & Control Approach	Organization: Independent oversight authorities for decision right assignment and controls (Quality Function) Process: Strict adherence to documented control processes, high degree of prescribed checks and balances placed on processes.	Organization: Cross-functional steering committee for broader decision controls, cross-functional teams for intra-team decision controls Process: Performance measures focused on results of process tied to team and individuals, with rewards and punishments levied in light of performance relative to goals or standards established.

Fig. 11–4 "QA/QC" Quality Management vs. Total Quality Management "Differentiating Factors."

how conformance to requirements is achieved. For a summary of the difference between "QA/QC" Quality Management and Total Quality Management, see Figure 11–4.

TOTAL QUALITY MANAGEMENT'S APPROACH TO KNOWLEDGE AND CONTROL

The core of TQM's ability to create value lies in its power to bring about an efficient creation and utilization of valuable specific knowledge at all levels of the hierarchy, while using more results-oriented control methods. While others, such as Drucker, Senge, and Garvin, have begun to a place a great deal of emphasis on knowledge and learning organizations, Jensen and Wruck appear to be the first to have developed a working model capable of addressing knowledge and control, a model that is shown to exist in practice.

The management techniques that fall under the umbrella of TQM can be viewed as a new methodology for organizing a company, representing a set of nonhierarchical, nonmarket-oriented organizing principles designed to achieve efficiency. TQM is described as nonhierarchical because its effective implementation disables many nonproductive aspects of the previous corporate hierarchy. TQM is described as nonmarket oriented because it does not use prices or formal exchange mechanisms, such as transfer pricing systems, to motivate cooperation or the transfer of decision rights. Efficiency is defined as providing customers the product and quality demanded at the lowest price while covering costs, including capital costs.

Two basic organizing principles are critical to the success of TQM, and understanding them lends insight into the principles governing the effective management of organizations. First, a TQM organization systematizes the scientific method in everyday decision making, which in turn leads to more effective creation and utilization of knowledge. Training employees in the scientific method improves their ability to organize and process data and to analyze problems, which in turn leads to more effective creation and utilization of knowledge. Second, a TQM organization adopts a new approach to internal control by making major changes in three types of internal systems: (1) its system for allocating decision making rights, (2) its performance measurement system, and (3) its system of organizational rewards and punishment.

SPECIFIC KNOWLEDGE AND THE DECISION RIGHTS ASSIGNMENT PROBLEM

Perhaps the most difficult problem facing an organization implementing TQM is how best to change the organization's system for assigning decision-making rights. Superficially it seems, both from observing companies and from the writings of quality experts, that TQM's primary message is one of massive and largely indiscriminate decentralization. More careful analysis and observation reveals that TQM's overall message is more subtle. Creating value through better use of specific knowledge is an important message of the quality movement. However, decentralization is the efficient approach to creating value only when individuals at lower levels both possess (or can create) the relevant specific knowledge at a low cost and can be motivated to use their decision rights to further the organization's objective. Otherwise, it is efficient to elect more centralized decision making.

It follows that equipping employees with skills to apply the scientific method and encouraging them to use their specific knowledge , while necessary, is not sufficient to create value. Successful TQM companies establish a system for assigning decision rights that curbs managers' tendency to over-centralize, yet at the same time establishes a structure and controls to guard against excessive decentralization. TQM organizations typically use a decision-making structure with steering committees (consisting of cross functional representatives) and process improvement teams. The steering committees help to prevent excessive decentralization by qualifying problem areas for study, separating out the significant from the trivial. The steering committees prevent excessive centralization by reassigning decision-making capability to the process improvement team from the existing hierarchy.

CONTROLS THROUGH PERFORMANCE MEASUREMENT AND REWARDS SYSTEMS

To this point, we have discussed how TQM programs systematize the scientific method, encourage the creation and utilization of valuable specific knowledge, and require a supporting system for allocating decision rights to be effective. Here, we explore the role of performance measurement and reward and punishment systems play in establishing more results-oriented internal controls. As noted earlier, effectively implementing TQM requires changing the system for allocating decision-making rights, generally in the direction of greater delegation. Potentially, this creates large agency or control problems; under TQM, employees with greater decision-making rights can use those rights to serve their own, rather than the organization's, interest.

Once an appropriate allocation of decision rights is established, organizations use both performance measurement and reward and punishment systems to help reduce remaining conflicts of interest. Without such changes the organization is likely to drift back towards its old way of doing things. Thus we predict that to make lasting improvements in efficiency, a company must make changes in both its performance measurement and reward and punishment systems. In particular, since TQM involves an extensive delegation of decision rights, we expect an increased emphasis on developing and monitoring new, more detailed, measures of performance. We also expect a strengthening of the association between rewards of all types and organizational performance.

To encourage individuals to change their behavior, management must develop and communicate new measures of success. In analytical terms, management's challenge is to redefine the objective of the organization and help employees understand how their actions help or hinder the new objective. The new performance metrics associated with TQM further this cause by establishing measures that define good performance and set performance improvement objectives. TQM performance measures emphasize close monitoring of the efficiency of the company's processes. Some of the more commonly used measures include manufacturing cycle time, product failure rates, on-time delivery rates, order lead times, new product development times, customer satisfaction indices, and waste or scrap rates. Performance measures applicable to individuals are sometimes couched in pay for knowledge plans, where compensation is tied to the acquisition of skills that are valuable to the company by increasing an individual's wages as he or she learns.

The new performance measures differ from traditional performance measures in three important ways. First, performance measurement systems in TQM organizations emphasize measuring productivity and quality from the standpoint of the customer. A customer-oriented approach to performance measurement helps prevent an organization from becoming inwardly focused and detached from its customers. Second, new performance measures are more operational rather than solely dollar-denominated or accounting-based measures. In most TQM organizations, such operational measures supplement, and sometimes even replace, traditional accounting measures such as product cost, labor rates, material or labor variances, and profitability. Finally and relatedly, new performance measures tend to be more disaggregate and function- or task-specific than traditional measures. Because they are more disaggregate and specific, they isolate the contribution of particular processes, functions, or tasks to the organizationwide quality effort. This helps employees focus on actions they can take to improve the company's performance.

Performance measurement systems in TQM setting also use a different approach to setting standards than traditional systems. Whereas traditional standards focus on average overall performance (for example, the average or overall percent of products without defects, or the average unit product cost), TQM standards are expressed in terms of both mean and variance. Many TQM organizations also "benchmark," comparing their performance to data available on the performance of peer or competitor companies.

To create lasting improvements in organizational efficiency, it is not sufficient simply to set a new objective and measure performance in a new way. In these circumstances, the behavioral effect is fleeting, attributed mostly to Hawthorn-related phenomena. In order to affect behaviors, incentives or reinforcements, positive and negative, must tie to group and individual performance, i.e., good performance must be associated with rewards, and weak or poor performance with a lack of rewards or punishment. Rewards can be monetary, such as raises, bonuses, profit-sharing plans, equity-ownership plans, or can be nonpecuniary, such as the utility generated by the participation process, the benefit of making one's job easier and/or safer, or public recognition. All these rewards are valued to varying degrees by employees, and therefore provide motivation or incentives. Furthermore, as many economists would agree, organizations do tend to benefit when they actively work to instill meritocracy, not egalitarianism.

Tying rewards to performance sounds simple and sensible, yet compensation, in particular monetary compensation, is one of the most hotly debated topics in TQM. Quality experts' views on rewards are important because many companies adopting a particular expert's quality management prescription also adopt his or her approach to compensation. For example, a company following both Deming's and Crosby's approach would shun all forms of incentive compensation for its employees. An analysis of the quality experts' case against monetary rewards leads one to conclude that their arguments are founded on a fundamental misunderstanding. Experts opposed to monetary rewards correctly recognize that a poorly designed incentive compensation system can destroy value because, in effect, it pays people to do the wrong things. From this potential problem, they incorrectly conclude that all forms of incentive compensation should be abandoned under all circumstances. In fact, it is important to recognize that *de facto* every organization, through both implicit and explicit contracts, has an incentive compensation system that includes nonmonetary and monetary rewards.

The fundamental issue is whether and how the incentive system associates rewards with contributions to company value. The problem is not that money is a "lousy" motivator, but rather that it is an extremely powerful motivator—perhaps too powerful in some circumstances. A poorly designed monetary reward system can easily worsen performance because it rewards employees for value-destroying behavior. Changing a company's allocation of decision-making rights and performance measurement system with no change in the structure of the monetary rewards system is likely to backfire. In essence, quality management is focused on the ability of the company to make collaborative, fact-based decisions that serve to meet the needs of the customer. The numerous standards for quality all prescribe various means for ensuring that decisions are indeed factually based. Equipment that is calibrated provides accurate data necessary to qualify or disqualify some test or inspection. Procedures, drawings, and other knowledge preservation devices which are controlled and kept up-to-date, provide factual evidence necessary to fabricate parts, inspect parts, devise inspection plans, provide the expected level of service, etc. However, all of these knowledge preservation tools are rendered ineffective if internal controls, in terms of performance measures and rewards and punishments, are not in place to reinforce the behavior expected. Ask any operator who is being reinforced by a supervisor to do work that is in conflict with the specification. The operator will most often do what he or she is reinforced to do, either through extrinsic or intrinsic reward or punishment. The same is true for the relationship between the president of a company and his or her board of directors.

Clearly, managers must take great care in designing and implementing the details of any incentive plan. It is the details of these plans that determine the difference between success and failure. See Figure 11–5 for a summary of viewpoints addressing monetary rewards.

OPPONENTS		
Group	**Indivduals**	**Rationale**
Quality Experts	• K. Ishikawa • P. Crosby • W. Deming	• Money is a poor motivator • Numerical performance measures are flawed and shouldn't be used
Behaviorists / Psychologists	• Alfie Kohn, et al.	• Rewards encourage people to focus narrowly on a task, to do it as quickly as possible, and to take few risks • Extrinsic rewards can erode intrinsic interest • People come to see themselves as being controlled by a reward
ADVOCATES		
Quality Experts	• J. Juran • S. Mizuno	• The reward system system (monetary, promotions) not only serves its basic purpose of rewarding human performance; it also serves to inform all concerned of the upper managers' priorities • If goals are revised but the reward system is not, the result as viewed by subordinates is conflicting signals. Most subordinates resolve this conflict by following the priorities indicated by the reward system
Financial Economists & Incentive Compensation Experts	• Michael Jensen, Harvard Business School • G. Bennett Stewart, Stern Stewart • Milton Friedman, Nobel Prize-Economics, 1976	• It is abundantly clear, from substantial marketplace evidence, that significant at-risk compensation strongly incentivizes behavior along the dimensions being rewarded • There is no precedent in all of society to suggest that humans continue to respond favorably to leveled outcomes for everyone

Fig. 11–5 Advocates and Opponents of Monetary Incentives.

INDUSTRY PERSPECTIVES ON QUALITY MANAGEMENT PRACTICES

As mentioned before, and as evidenced in Figure 11–6, there is not a singular approach to quality management in the oil industry. Figure 11–6 provides a summary of information pertaining to quality management approaches in a sampling of some of the larger domestic oil companies. While the identities are not given, it is clear that both significantly different and strikingly similar practices exist across the industry. Most importantly, there are some trends indicated by the research, including:

- Increased emphasis on organizing technologies such as TQM for creating and using specific knowledge
- Increased emphasis on controls through perfor-

QUALITY MANAGEMENT APPROACH	INDUSTRY TENDENCY
• Control By Process	◕
• Control By Results	◑
• Influence By Gurus	◑
• ISO Certification Upstream	◔
• ISO Certification Downstream	◕
• MBNQA Application	◔
• MBNQA Criteria Worthwhile	●
• Formal QA Group	◕
• Formal TQM Group	◔
• Integrate ESH With Quality	◔
• TQM In Future	◕
• Financial Focus	◑

Fig. 11–6 Recent Trends.

mance measurements and rewards tied to results vs. strict controls through process/independent oversight

- Increased efforts on the part of the standards community to move in the direction of continuous improvement/TQM
- Increased emphasis on more directly linking operational performance measures to financial performance
- Increased emphasis on integrating new requirements, such as environment and safety, into the control system versus creating new oversight organizations.

RESULTS OF QUALITY MANAGEMENT

s noted in the introduction to this chapter, considerable confusion surrounds the notion of quality management. This confusion, coupled with the lack of a consistently applied scientific method tied to quality management, makes assessing the results of quality management problematic. Some studies and articles have appeared which attempt to assign success or failure to the causal factor of quality management, including writings from both academia and industry. These more rigorous efforts find anywhere from a negligible to positive effect associated with the implementation of quality management practices. In addition, there are countless anecdotes addressing the influence of quality management practices on success of the company, with little conclusive evidence provided in the aggregate. Lastly, there seems to be little attempt to capture the costs and benefits accruing to quality management practices, with exception given to those companies rigorously tracking a cost of quality measure, which further contributes to the dilemma of assigning causality.

Booz·Allen's Richard Spitzer has taken an inductive approach by looking at the quality management practices of companies which have experienced significant, abnormal shareholder returns. His hypothesis is that truly useful quality management practices should be reflected in shareholder value. Spitzer uses the Stern Stewart Performance 1,000, a corporate ranking based on shareholder value creation, to establish the benchmark companies, and then proceeds to highlight the quality management practices prevalent in these companies. Spitzer's findings are that TQM practices of cross-functional process teams and steering committees, controls through performance measures and rewards focused on results (including customer satisfaction), and delegation of decision rights to the bearer of specific knowledge, are correlated with companies achieving significant shareholder value creation.

Numerous other approaches have attempted to assign success or failure to quality management, including many pertaining to the energy industry. For example, in Alberta, Canada, major customers are encouraging their suppliers to join them in adopting TQM, and the Petroleum Services Association of Canada is helping its members to learn about the process. Some success stories are emerging, but Alberta's service sector has been slow to respond. Many say they are interested, but claim to be too busy to act on it.

Diamond Shamrock points to quality management for success as well. In the past several years, Diamond Shamrock, the regional refiner and marketer based in San Antonio, has established itself as one of the major players in the high-stakes petroleum markets of the Southwest, Mid-Continent, and lower Rockies. Diamond Shamrock has prospered in this hotly competitive environment with a strong mix of modem retail gasoline/C-store outlets in key markets and an efficient network of refineries, pipelines, storage, and distribution facilities that is well-positioned to serve its retail, wholesale, and commercial customers. Perhaps the single most important ingredient in the company's success, however, is its evolving corporate culture. What's involved here, in the words of chairman and CEO Roger R. Hemminghaus, is "a renewed commitment to customers, operational excellence and productivity improvement through total quality management." In Hemminghaus' view, TQM is not just another management program, but a permanent reshaping of the company's management and operating style; a process that gets much more than just lip service from corporate higher-ups. According to Hemminghaus, TQM is an active and ongoing process that eventually will train and involve every one of Diamond Shamrock's 5,000-plus employees in improving productivity, teamwork, and customer service.

Results particular to ISO 9000 seem to be fairly positive in some circumstances, particularly when certification allowed market opportunities to be gained or at least maintained. Indeed, some directly attribute cost savings attributed to the controls established through ISO 9000, although not much reporting is provided concerning the costs to erect and maintain such paper-driven systems. DuPont also points to financial success associated with the costs reduced, but does not provide much information pertaining to the costs required to build, operate, and maintain such a system. Others have reported not so favorable experiences concerning ISO 9000, raising significant concerns regarding the cost and productivity impact concern procedure driven approaches to control.

Quality management is not easy to implement, and many companies find that the programs get "stuck" and fail to generate the expected benefits. The fact that for some quality management requires huge up-front investment in employees, as well as necessitating an entirely new approach to internal control, sheds some light on why implementing quality management is so difficult. Of course, the excuse of failure to properly implement tends to be the scapegoat for numerous management fads of the time, and therefore does not shield the concept from rigors of the marketplace.

The absence of more rigorous empirical analysis of the impact of TQM on company performance may be due to the absence of data: management seldom publicly announces the adoption of new management approaches when they occur, perhaps due to concern that such an announcement would be interpreted by the market as a signal that the old ways had failed. In addition, TQM often is considered only when a crisis is looming, so an announcement that TQM is to be implemented might signal that management foresees a crisis. In fact, of the 17 Baldrige winners thus far, we are aware of only one, Marlow Industries of Dallas, Texas, that claims to have adopted TQM strategies when the outlook for corporate performance was positive.

THE FUTURE OF QUALITY MANAGEMENT

QUALITY MANAGEMENT FOCUS: PROCESS VS. RESULTS

ompanies will continue to see various standardized approaches to quality management, some more prescriptive, more process control based, others less prescriptive and results or performance based. Regulated industries, at least in the near term, will likely continue to see some organizational oversight and process prescription to some extent, in the form of the more traditional QA/QC models of quality management; however, when performance measures and rewards and punishments have historically been absent, the control and oversight approach is oftentimes warranted. There are indications that the direction of quality management practices is away from the internal control model of strict compliance to procedures and separate quality organizations, towards internal control models such as what Jensen proposes.

Companies are relying more on a results-oriented approach, where priorities for quality are built into the internal controls of performance measurement and reward and punishment. Placing the emphasis on building priorities for quality in these truly behavior-driving processes of planning, goal setting, performance measurement, reward, and punishment, is proving to be more effective than the procedure-driven approach associated with many standards for quality management. In fact, some Baldrige Award winners and others have bemoaned the failure of draconian approaches to achieve control through proliferation of documents, which rightly or wrongly, tends to be the way of life for adhering to the QA/QC quality management standards.

Innovations in organization design which tend to instill marketplace accountability are also proving to be effective in realizing quality results without the bureaucracy of compliance typically found with certain industry quality standards. Organizational forms, such as business teams, asset management teams, self-directed work teams, horizontal organizations, virtual corporations, and market-based performance measures, such as economic value added, market value added, may become the chosen course of action vs. strict adherence to industry quality standards and continued reliance on purely functional organizations.

QUALITY MANAGEMENT SCOPE: REGULATORY COMPLIANCE PROGRAMS

The areas of environment, health, and safety are gaining increasing prominence in the 1990s as a result of laws and regulations coming into effect. These areas affect oil companies as well, particularly the downstream refining and petrochemical businesses. For example, the Risk Management Prevention Program in California, the Clean Air Act of 1990, and OSHA 1910.119 Process Safety Management all include programs for the management of process safety. OSHA 1910 is a new specification which will affect the refining and production side of the oil business.

A number of professional organizations have prepared process safety management programs. These include the American Petroleum Institute, the Center for Chemical Process Safety, and the Chemical Manufacturers

Association. Two key features characterize all of these programs. First, they are provided as management tools, intended to describe ways in which plant management can organize and control risk and reliability. The second key feature is that they are generally nonprescriptive, meaning the regulation or standard does not specify exactly what has to be done. It simply states that the plant should be managed in such a way as to ensure its safety. For example, the regulations do not require each plant operator to have, for instance, 40 hours of training. Instead, they require that the operators be properly trained so that they can carry out their work safely.

Related to quality management, there seems to be an effort to lump the programmatic approaches intended to address these three areas (environment, health, and safety) into a new version of risk management. Not to be confused with the more familiar concepts of risk management which either focus on managing financial risk or are related to insurance, these new approaches are focusing on assessing and mitigating operational risk. Here, risk is generally focused on the incident probability and related consequences to environment and health. And, as one might suspect, the standards organizations are not too far behind and are already in the process of including these concepts in future updates to their respective quality program standards, including ISO 9000. In fact, one could argue, that insofar as quality programs are intended to ensure conformance to requirements, then these requirements addressing environment, health, and safety would logically be included. Bruce Barge notes:

> *"Quality management and risk management are two sides of the same coin...Quality management is an interdisciplinary approach to business operations. It looks at processes that cut across the entire organization. It sees all the outcomes and results of an organization—productivity, the quality of products and the losses that occur within processes—as products of the same organizational processes. Quality management demands that each risk manager ask: how can we incorporate the loss control perspective into process improvement? The question, by implication, requires the risk manager to understand the entire organization, and how loss control enhances the organization's quality management and contributes to the organization's ultimate success. Recent winners of the Malcolm Baldridge National Quality Award—including IBM-Rochester and Federal Express—report that they markedly improved their loss control results as part of their quality efforts. Quality is the flip side of risk. And maximizing quality means minimizing the chance of expensive corporate losses."* [1]

From a TQM perspective, the risk assessment tools, such as fault tree analysis and HAZOPs, are yet another set of analytical techniques to be used during the course of applying the scientific method referred earlier to problem solving.

CONCLUSION

he bottom line is that for quality programs to succeed and deliver results, they must ensure that the cost and benefits of quality be rigorously measured and that a commitment to quality become engrained in the company's culture.

---

1 Bruce Barge. "Total Quality: A Risk Management Opportunity," *National Underwriter Property & Casualty Risk & Benefits Management*, February 17, 1992, Issue N7, p. 17.

CHAPTER 12

HUMAN RESOURCES DEVELOPMENT

A business perspective on the oil and gas industry would be incomplete without discussing the critical human factor. Too often, due to its capital intensity, the oil and gas sector is associated first and foremost with massive assets and technology. The relevance and complexity of human resource development is often underestimated.

This chapter explicitly addresses how human resource development can make a considerable contribution to the business.

This chapter is structured in four sections:

- The first section gives an historic perspective on Human Resource Development (HRD) in the oil and gas sector
- The second discusses human resource development and puts it at the core of the human resource function. Although human resource development typically represent only a fraction of the human resource function, it also has, in comparison to the other HR activities, probably the largest potential to make an impact on the bottom line. Increasingly, HRD is needed in order to have the right staff available at the right moment in the right place. Doing this in a constantly changing, global business environment for a large multinational workforce is no easy task.
- The third discusses a range of developments that are likely to influence HRD policies in the near future. The implication here is that companies that anticipate these developments and react well may turn what others perceive as threats or problems into competitive opportunities.
- The last section makes the point that the development of human resources has to fit within a company's past, its culture, and current constraints. There is no such thing as a "best" HRD program. Also, when it comes to implementing and changing deeply rooted human resource practices, one will have to realize that patience, dedication and realistic expectations are called for.

EVOLUTION OF HUMAN RESOURCE DEVELOPMENT IN THE OIL & GAS SECTOR

ver the last 40 years, the oil industry has changed dramatically. The massive growth of demand was the underlying driver of most of the change. Downstream demand has grown worldwide, leading to significant investments in

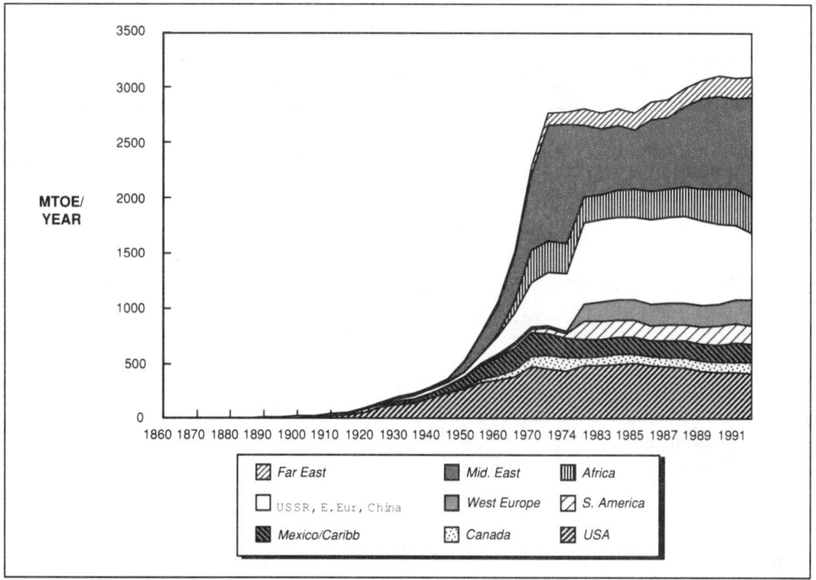

Fig. 12–1 World Crude Production.

refining capacity on all continents. Increasingly demanding specifications of the end product have increased the need for more complex processes and higher capital expenditures.

Figure 12–1 shows how this global demand has increasingly been met by production from outside the United States. New producing provinces, such as the Middle East, have grown in importance. Most of these provinces were initially operated by U.S. or European oil companies. Today, national oil companies control the majority of production.

These changes have had their effects on human resource management in the oil industry over the last 40 years:

- An increase of international operating activity, especially upstream, has forced oil companies to improve capabilities to staff overseas projects. Although progress has been made, many oil companies recognize that they are still far away from the desired human resource procedures for foreign operations. Typical obstacles include: lack of understanding of foreign operating environments, reluctance to change, and insufficient return positions available for expatriates who return to the home country, among others.

DEPARTMENTS	LEVEL AT WHICH DEPARTMENT OPERATES	
	CORPORATE/CENTRAL	OPERATING COMPANIES /DECENTRAL
- Labor Relations	✔	●
- Coordination Professional Communities	✔	✔
- Studies	✔	✔ ●
- IT	✔	✔ ●
- Personnel Administration	✔	✔ ●
- Human Resource Development	✔	●
- Recruitment	✔	●
- Training (Strategic, Tactical, and Executional Levels)	✔	✔ ●
- Compensation	✔	✔ ●
- Retirement	✔	●
- Expatriate Personnel	✔	✔

LEGEND

✔ = For staff that has been awarded a high potential. This category typically includes internationally mobile staff

● = For staff that has not been awarded a high potential. This category typically includes local, nonmobile staff, working in either the domestic or the foreign subsidiaries.

Fig. 12–2 Typical Breakdown of the Human Resource Function.

● The enormous growth in size of the oil companies has reduced the number of "all-round" jobs and increased the number of specialist jobs. In all professional fields, oil companies employ today highly trained staff: technical, financial, legal, managerial, etc. The top jobs, once taken by entrepreneurial individuals in a double role as owner/director, now go to professionals who climbed the career ladder. Human resource departments have a major influence on the developments of these careers

● Increased technical, financial, political, and geographical complexity of the industry have boosted the need for top staff. Exploration and production in deeper water, harsh climatic conditions, increasingly complicated process technology, complex tax structures, environmental sensitivities, host government issues, and many other factors have made the potential cost of a mistake enormous. Figure 12–2 illustrates the need to match an individual's abilities with the demands of a job! The human resource department plays a main role in this process.

HUMAN RESOURCE DEVELOPMENT AS THE CORE OF THE MODERN HUMAN RESOURCE FUNCTION

he traditional human resource function, often referred to as a personnel function, has a departmental breakdown as shown in the left column of Figure 12–2. This model was originally designed for administrative tasks in a one-country operation, but has gradually been adjusted in an attempt to cope with changed circumstances and new demands. Often structural changes have not been made. For instance, a department for expatriate staff was added, without achieving true integration of expatriation issues with other human resource or company issues. Lacking a structural rethinking of the function and its key processes, a suboptimal situation will occur.

Today's human resource function is designed for its current tasks, and for its internal customers: the line function and the employees, each with their own demands and expectations. High-potential staff, working worldwide, require close attention from the human resource function at the central and local level. Tailor-made processes are required to meet these demands. Unfortunately, however, few companies have structurally rethought their human resource function and many are confronted with a human resource function that is simply not fit for the purpose. Problems encountered include:

- Confusion of the means and the goals of the human resource function, resulting in a lack of goal orientation.
- Inefficiencies due to communication and coordination have the potential to have disproportionately negative effects. Apart from the financial cost, they can have a strong negative effect on staff motivation and the credibility of the company. To many staff, the human resource function is an important part of the face of the company.
- Decision power with operational implications

rests with the human resource professionals who, due to a lack of operational experience, are often not qualified for such decision making.

- Overemphasis on administration and incapacity to truly support the business strategy.
- Internal customers, i.e., the line functions like exploration and production, refining, and marine as well as the employees themselves, are not served as well as they could and should be.

The primary goal of the modern human resource function is to support the execution of the business strategy. Administration has become a means rather than a goal. In this model the management and development of staff have become a top priority. Although differences between companies exist, the driving principle remains the same: top quality staff members are rare, and their use and development should be optimized at all levels of the company and not just at divisional or operating level.

When discussing HRD, it is useful to distinguish between two different groups of staff, each requiring different types of attention (see Fig. 12–2):

- High-potential staff, including the internationally mobile staff. These people are typically managed, monitored, and developed with a coordinating central function
- Local staff that is locally recruited, employed, managed, and monitored

This chapter focuses on the development of high-potential staff, those employees who could rise to the senior ranks of management. This task represents only a fraction of the total human resource activities, but it is at the same time the part of the human resource function with the longest levers of corporate performance. Human resource development should be, but rarely is, the core task of the human resource function. Human resource development is built around six important processes:

- Planning
- Recruitment
- Training
- Job rotation and transfers
- Appraisals and promotion
- Compensation

Each of these six core processes will be discussed and a chart will show how practices in key aspects of each process range from novice to world class performance.

PLANNING

In order for an HRD program to provide a good service to its internal customers, an HRD planning process is required that hangs closely together with the business strategy, as shown by Figure 12–3. The internal customers are the line functions such as refining, exploration and production, marketing, and sales, who have endorsed the business strategy. Only when that connection exists can a meaningful and effective HRD program be executed.

The business strategy will indicate, amongst others, the (new) capabilities the company needs for a successful implementation of its strategy. The human resource plan should translate the capability requirements into people requirements. As soon as a company needs to change capabilities, there is a direct impact on the people side, be it through recruitment, training, transfers or whatever. Capabilities reside, at least partly, in the staff working for the company. Obviously, a planning process

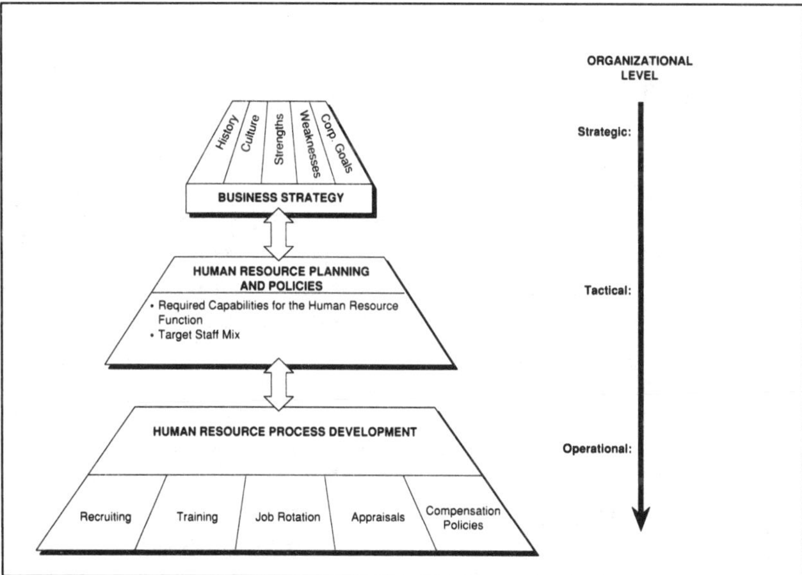

Fig. 12–3 Integration of Human Resources Development with the Business Strategy.

should be realistic and recognize its limitations. Unanticipated major discoveries economic and political discontinuities and the introduction of radical new technologies—and the implications of these events for HRD—are difficult to predict.

Practices in HR planning and their linkage between business strategy and HR policy are shown in figure 12–4. The left column (Level 1) indicates novice or baseline performance while the right column suggests practices of a world class high-performance organization

RECRUITMENT

An important output of the human resource plan is the mix of new staff to be recruited. Recruitment has two targets:

- Quantitative targets, which require insight into turnover, retirement, net growth of demand by the company, host government demands, split between centrally managed staff and locally managed staff, contractor policies, etc.
- Qualitative targets that require a view on the type of jobs that will need to be filled in the future in terms of disciplines, skills, languages, nationalities, personality, potential, mobility, required training/preparation, "country-specific entry requirements," etc.

The decision of where and how to recruit is becoming increasingly important and less evident. The number of graduates with petroleum-related degrees is falling in most western countries; the competition for high-quality staff is increasing; some companies may decide to have a program to recruit older people with relevant experience; recruitment of (specialist) contractors may become increasingly important to some companies. The actual recruitment is best served by a mixture of line staff and HR professionals. A candidate's professional strength and personality can best be tested by those who have been in the field themselves. Also these people are better qualified to convince the candidate to join on the basis of a true picture of the job.

A recruitment plan must also reflect the company's view on cultural diversity. Few oil companies have a truly global approach towards human resources with senior positions in a country being filled by a variety of nationalities. Shell and ABB (as a nonoil example) regard

	NOVICE → WORLD CLASS				
	Level 1	Level 2	Level 3	Level 4	Best Practices
HR Strategy	Respond to user requests—no forward looking strategy	Independent of business strategy	Reactively tries to support top line business strategy	Forward looking, driven by requirements of business strategy	Proactive with HR issues integrated into business strategy development
Plan for HR Strategy Implementation	No formalized plan	Specific objectives included in plan	Specific objectives and key performance indicators included in plan	Specific objectives, key performance indicators, activities and timing included in plan	Specific objectives, key performance indicators, activities, timing, costs and accountability included in plan Prioritization of initiatives
Policies	Outdated policies not supportive of current business needs	Incomplete or ambiguous policies	Policies serving the HR function rather than business needs	Policies consistent with the business and HR strategies	HR is a leader in establishing people policies that support the business and HR strategies
Strategy, Plan, and Policy Communication	Limited communication	Communicate policies	Communicate policies and plan	Communicate policies, plan and strategy	Communicate policies, plan, strategy, vision mission and values; and encourage two way dialogue with employees

Fig. 12–4 HR Strategic Planning Practices Ranked From Novice to world class. Sources: BA&H analysis, Best Practices Database.

cultural diversity as a potentially rich source of innovation and competitiveness. Figure 12–5 shows the four typical phases of globalization of a human resource function and the current position of several oil companies.

The recruitment practices are outlined in figure 12–6 demonstrate a substantial range of effort and effectiveness which may be invested in staff sourcing. It is important to recognize that achieving "world class" recruiting performance may not be appropriate for all oil companies (e.g. those downsizing or anticipating low growth).

TRAINING

A training program should be put together in connection with the human resource development plan. Some companies, especially those with a view on long-term staff development, have every recruit go through an initial training program. These programs are often run at a company training center. Part of the training is often given by high potential line staff. The multiple advantages are:

- Required technical skills are developed.
- Staff is prepared to handle uncertainty and change.
- Team-based performance can be stressed.
- The corporate infrastructure can be explained.
- A network starts to be developed.
- Standards are set
- Knowledge is distributed.
- A common terminology is adopted.
- Performance can be observed in a low-risk environment.
- Expatriation can more easily be justified towards a host government on the basis of the training received.

When operating in less-developed countries, an important capability for oil companies is to offer host country staff the possibility to get top level training.

Training practices covering a spectrum from none to advanced to dynamic individually tailored feedback driven programs are characterized in figure 12–7. Clearly the training program capability should vary with the company's need for building skills and the employee's potential for impact on the company.

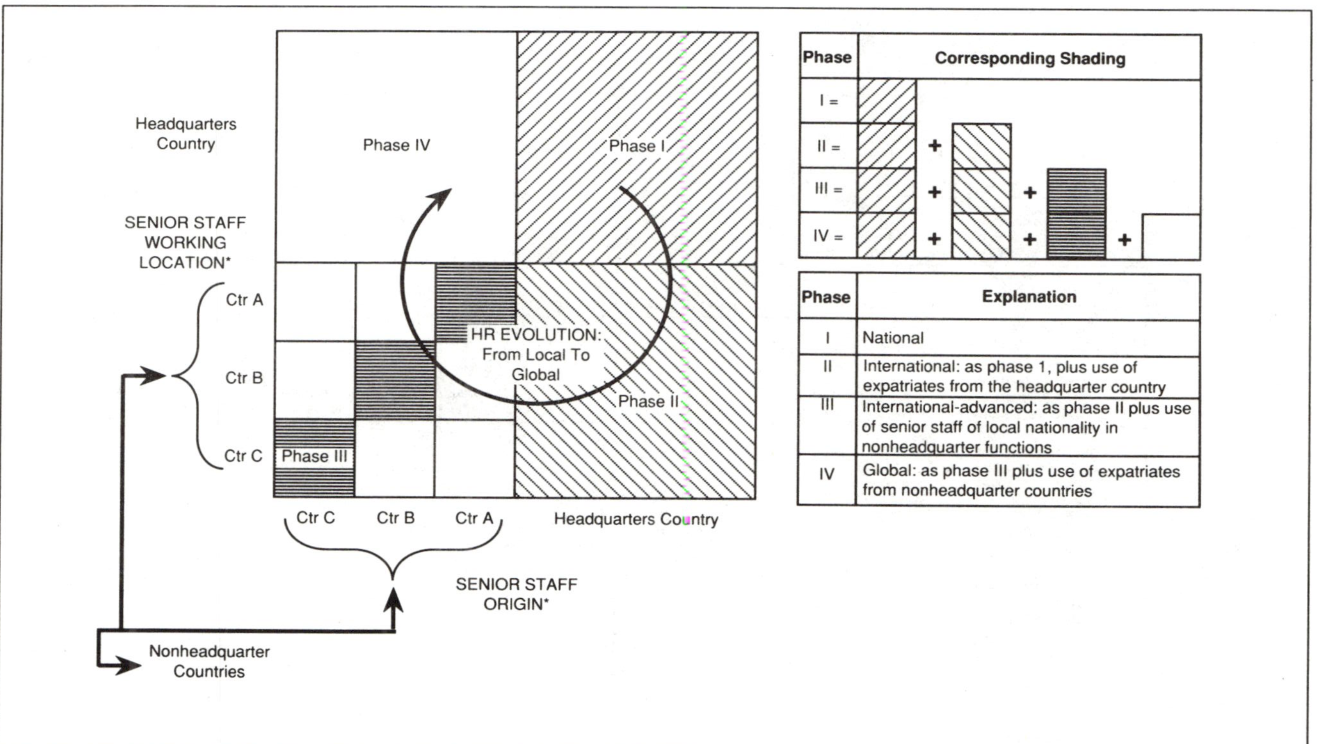

Fig. 12-5 The Four Phases of a Globalizing Human Resource Function.

	NOVICE → WORLD CLASS				
	Level 1	Level 2	Level 3	Level 4	Best Practices
Candidate Sourcing and Screening	Ad in campus and/or industry publication with only limited identification of desired qualifications	Use advertising, resume books, and/or executive search firms	Use advertising, resume books, executive search firms, write-ins, and/or employee referrals with clear identification of short term qualifications	Supplement Level 3 with participation in on campus recruiting	Supplement Level 4 with targeted communications through participation in campus events Tap into professional associations and alumni networks Clearly define desired qualifications for successful long term career
Interviewing / Selection	Untrained interviewers Interview based exclusively on resume	Trained interviewers Good definition of skills required for the position	Interview team with appropriate background to assess candidate	Interview team with appropriate background to assess candidate Additional criteria targeted at assessing long term potential	Interviewers all portray consistent image of the company Process appropriate to the level and amount of information required (may include: case solving, presentation, testing, etc.)
Offer, Feedback, Closing, Selling	Lengthy delays in response Contingent offers	Impersonal, formalized response	Fast response	Personalized response Fast response	Immediate, personal response Develop goodwill regardless of decision Continue to sell the company after offer
Cross Organization Coordination	Duplication of effort	No communication	Independent approach for some units or geographic areas	Centrally coordinated strategy	"One firm" approach with continuity of recruiting staff
Relationships with Sources	Contact sources as needed with no continuity or long term relationship	Relationships only if initiated by source	Relationships with conveniently located sources	Relationships with well defined core sources	Continuous presence with one face approach at core sources
Nationality Seeking / Global Recruiting	Nationals recruited for wage positions	Nationals hired up to mid-level positions with little chance of further advancement	Regional staff secured with potential to run local operations	International recruiting in geographies where current operations are located with potential to be staffed anywhere in organization	Global recruiting from core schools with purpose of developing international managers

Fig. 12–6 Recruiting Practices. Sources: BA&H analysis, Best Practices Database.

	NOVICE → WORLD CLASS				
	Level 1	Level 2	Level 3	Level 4	Best Practices
Orientation	None	Explain rules	Include overview of job, process, site, etc.	Level 3 plus company vision, goals, and values	Level 4 plus commitment and team building exercises
Needs Identification	No formal programs or tools Identified by individuals	Limited to skills necessary to perform current job	Standardized by job level—skills necessary to perform current job and move vertically within function	Coordinated with HR development process	Individually tailored—with forward looking emphasis matching organizational goals
Program and Structure	Not defined—only on an as needed basis	Some on the job training	Standardized program combining formal and on the job training	Flexible program including job rotation and geographical mobility	Dynamic, forward looking training, content and delivery method adjust according to need
Tracking	Non-existent	Inconsistent tracking	Tracking a local level for legal purposes	The provision of training and its effectiveness are tracked company-wide through HRIS Information is accessible only to HR organization	The provision of training and its effectiveness are tracked company-wide through HRIS Information accessible directly to appropriate end-users
Course Feedback and Results Measurement	None	Keep track of attendance and number of courses	Level 2 plus evaluation of course content and presentation	Level 3 plus feedback from attendees concerning effectiveness of training in HR development	Training related to improvements in key performance indicators (i.e., productivity, quality, etc.) Continuous efforts to improve the effectiveness of training

Fig. 12–7 Training Practices. Sources: BA&H analysis, Best Practices Database.

JOB ROTATION AND TRANSFERS

A job rotation system is an essential part of a human resource development program. It has the potential to deliver high value because it:

- Provides employees with the largest possible pool of training opportunities in terms of culture, type of job, and variety. Future senior managers can in this way be developed, tested, and prepared in their early careers.
- Provides projects with the appropriate technical, commercial, and managerial staff to make up for shortages of local expertise. The best individual can be mobilized for a certain (critical) job.
- Ensures that lessons learnt in one location are shared throughout the company. At the same time it ensures that new thinking and approaches are developed for old problems. Established practices will be challenged and improved.
- Helps keeping the company together by creating a human network and shared culture.
- Instills an atmosphere in a nonthreatening way where change is the rule rather than an exception.

For job rotation to be successful, at least two conditions have to be met. Firstly, a sophisticated system is required that profiles the company's jobs and the employees in terms of skills, needs, level, etc. This system is needed to match "demand," i.e., the job, with "supply," i.e., the employee. This matching is especially complicated due to the domino effect: if one employee moves to another job and/or location, the job he or she leaves behind often will have to be filled by a replacement. The job profile process has a clear interface with the overall human resource plan. Secondly, the company culture must fully support job rotation. Bosses need to accept that employees have to be trained and coached, while at the same time accepting that in the near future they may either see these employees transferred to another job or find themselves in another job and country. The processes that support the job rotation system must be adhered to with discipline and professionalism in order to have comparability across the company, and across the world. Annual appraisals must identify developments needs, workloads need to be communicated to have the right staff available at the right moment, and so forth.

Staffing practices associated with effective job rotation are suggested in figure 12–8. Many companies are strong in some aspects of this

	NOVICE → WORLD CLASS				
	Level 1	Level 2	Level 3	Level 4	Best Practices
Posting Openings	No open posting, only through the grapevine	Open posting, but limited to job specific location Often a manual or paper system	Posting extended to other targeted areas of company Respond through HR who coordinates efforts between employees and hiring managers	Division or company wide posting Respond through HR who coordinates efforts between employees and hiring managers Often an on-line or automated telephone system	Posting company wide, including international locations Available for employees to apply directly to hiring manager based on established criteria On-line or automated telephone system
Candidate Identification and Communication Across Business Units	Identification locally without communication across business units	Identification by managers within a business unit (possible hoarding of talent exists)	Identification linked to performance and potential possibly through annual review but limited systems to facilitate transfers across business units	Identification linked to career development and long term personnel requirements with communication across business units	Level 4 supplemented with a fast track for high potential employees identified across whole company
Selection	Local manager makes selection	Local team makes selection	Local team makes selection with counsel from next management level	Local selection with linkage to strategy and personnel planning	Level 4 supplemented with explicit attention paid to high potential employees in order to assess career and personnel planning implications
Succession Planning and Development	No plan, look to current immediate senior reports	Successful line managers promoted to senior managers	Senior managers combat each other in a tournament fashion to reach the pinnacle	Potential of mid to high level managers assessed and high potentials are assigned to high profile developmental positions	Early identification of high potential employees, analysis of their performance in developmental positions, a well defined fast track to have qualified senior managers

Fig. 12–8 Job Rotation and Staffing Practices. Sources: BA&H analysis, Best Practices Database.

process (e.g. company wide postings) but weak in others (e.g. potential candidate identification).

APPRAISALS AND PROMOTION

Before discussing the development of high-potential staff, one has to define the pool. It is obvious that with an increased size of the pool, the costs will rise but also the likelihood that the true high potentials will be identified. Different companies strike the balance in different ways. Shell has a system in which the "International Staff" is observed during the first part of their careers until an ultimate estimated potential can be allocated. After that, there is a tiered system with the highest priority for the highest potential staff. This system has a wide coverage, but it is expensive.

Appraisals often address recent performance and less often the potential of an employee. Performance assessment should be done against a set of criteria known to the employee. The criteria applied are a function of a company's values and may, for instance, emphasize teamwork rather than individual performance, safety, cost awareness, quality of work, knowledge, security, service mindedness, initiative, communication, organization, and other traits. Another difference between companies is the number and level of people who have an input into the appraisal of an employee. Is it done by exclusively the direct boss, or are subordinates of the appraisee interviewed for their input, i.e., "upward appraisal," as well? Is there input from a central HR function or manager? Appraisal input and effort are clearly increasing with the level an employee reaches in the company. For an appraisal to be meaningful, it should have a true impact, for instance, by accelerating promotions, being linked to bonuses, and so on. U.S. companies typically have a more transparent and direct link between performance appraisals and career progress and financial rewards than European companies.

Potential appraisal can theoretically be separated from current performance, but it should not be. Potential appraisal focuses on the characteristics of the individual, such as analytic power, sense of realism, imagination, drive, and capacity to combine overview with sense of detail, among other traits. A good potential appraisal system allows companies to select and prepare people for the top jobs. The associated problems are, however, considerable. If you communicate the evaluation to the employee too early, you risk making mistakes and in doing so, damage the motivation and loyalty of the high-potential staff that have not yet been identified as such. However, if you communicate potential too late, you risk losing high-potential people that run out of patience and leave the company.

	NOVICE → WORLD CLASS				
	Level 1	Level 2	Level 3	Level 4	Best Practices
Performance Objectives	Explicit performance objectives not defined	Performance objectives defined Vague measurements	Performance objectives and measurements clearly defined No clear connection between objectives and organizational performance	Objectives aligned with corporate objectives, but line of sight is too far for most employees	A few, key measurable objectives established: aligned with performance drivers aligned with job design (team versus individual objectives)
Appraisal and Feedback	None or informal only, undifferentiated "check the boxes" cursory process	Top down appraisal, one on one with supervisor only, yearly	Objectives and measurements clearly defined, criteria aligned with group goals and job design No or unclear linkages to compensation, training or development	Substantial management time dedicated to process, peer and self evaluation included	360° feedback, jointly agreed upon goals, timing to match performance cycle, linked to compensation, training, and development
Development Needs Identification	Identified by individual	Standard development program	Development tailored to individual, drives next assignment	Driven by feedback program, training cataloged	Forward looking skill gap analysis between current employee level and corporate needs drives development programs
Career Planning	Career assessment tools and support are not widely disseminated Career paths are not widely known	Sample career paths identified	Level 2 and career assessment tools and counseling available to help individuals define career objectives and plans	Individual objectives translated into assignment plans driven by potential	Career planning tied to personnel planning, flexibility, career counseling
International Exposure	No programs, ad-hoc success	Suggested/required for advancement (later in career)	Exchange programs with specified duration	Encouraged early in career with rotations designed to further development, senior sponsor on "both ends"	Globally integrated assignments with participation of high potentials prerequisite to top management
Identifying High-Potential Employees	High potential = senior position, "hold on" to best young employees	Patchy identification, politicized process	Local managers look for potential, program to identify high potentials in place	Characteristics of high potentials defined, impact on career plan	Identified early ≤5 years by senior level, fast track defined to result in internal development of senior managers, senior level tracks career

Fig. 12–9 Appraisal and Career Planning Practices. Sources: BA&H analysis, Best Practices Database.

Once potential has been assessed, it should be reassessed every year without bias. If not, the danger of the self-fulfilling prophecy is very real: potential assessments start driving career plans and promotions, while performance is no longer fairly assessed. Comparing staff potential across the globe can only be done with a kind of quantified scale. This process suggests an accuracy that in reality cannot be achieved. The credibility of the system is therefore limited, and any overemphasis on the early allocation of an employee's potential should be avoided.

Practices for staff appraisal and career planning are arrayed in figure 12–9. HR plays a key role in maintaining quality and consistency in managers' application of the appraisal process across the company.

COMPENSATION

Although most people are motivated by factors other than money alone, underestimating the importance of compensation is unwise. An employee wants to be treated fairly and receive competitive compensation. Especially in the sometimes tough upstream business, where oil typically is found in the most desolate places, expatriate employees accept a foreign posting in return for something concrete.

The objectives of a fair compensation system should be:

1. To attract and retain staff of the desired caliber, and so
   - avoid unjustified differences
   - meet employees' needs
   - be transparent, while also maintaining the capacity for flexibility
   - reflect job weight and performance

2. To provide the right incentives, e.g., the proper mix between individual and group performance. U.S. companies typically emphasize the individual component more than most of their European competitors.

3. To provide the employee with a satisfactory lifestyle and an incentive to work abroad and move regularly. This may include local housing, specific hardship allowances, free education for children, and other benefits.

	NOVICE ⟶				WORLD CLASS
	Level 1	**Level 2**	**Level 3**	**Level 4**	**Best Practices**
Philosophy/Strategy	Focus on fairness and internal equity only	Focus on fairness, internal equity, and external competitiveness—"attract and retain"	Level 2 with focus on key organizational goals	Focus on key organizational goals—communicate, reward, and differentiate based on performance	Level 4, but as a total compensation package (including indirect pay such as benefit packages)
Salary Plan and Structure	Base salary with merit increases as the only widely used form of direct pay	Variable pay component, but only for senior management and sales organization	Variable pay component for all/most levels of the organization	Level 3 with linkage to performance management system	Variable pay component linked to performance management system Aligned with job design (balance of team pay, individual pay) Variable reward amount large enough to make a difference to employees but within their risk profile
Expatriate Compensation Policies	Case by case—"cut the best deal"	Based on host country salaries	Adjustments to home salary for cost of living changes, hardships, local housing and taxes	Level 3 plus well-defined adjustments reflecting the variety of nationalities and locations and an incentive to move abroad	Level 4 with a system for differentiating between international and national expatriates

Fig. 12–10 Compensation Practices. Sources: BA&H analysis, Best Practices Database.

Compensation practices are outlined in figure 12–10. Many of the companies which exhibit world class practices remain unsatisfied in dealing with equity issues in compensation levels for multi-national expatriates and managing complexity in designing performance based incentives.

FORCES OF CHANGE FOR HUMAN RESOURCE DEVELOPMENT IN THE OIL AND GAS SECTOR

In today's environment, so different from the booming seventies, a range of developments can be discerned with the potential to have a profound impact on the human resource process in the oil and gas sector. The most relevant ones are discussed here.

■ ***Cost is increasingly important.*** Cost pressures are likely to remain for many years due to the threat of overproduction hanging over the market. Most oil companies believe that oil prices will remain low for several years and are looking at ways to improve their cost structure. While reducing costs, companies ought to be careful not to alienate their workforce.

Although a company's internal salary costs are often modest, cost-cutting will lead to headcount reductions, and essentially, more work will have to be done by fewer staff. In doing so, the additional burden that is put on the remaining staff should not be underestimated. Already there are examples of companies that have simply "cut 20% across the board," and as a result cut too much in the high-value areas and too little in the low-value areas. The value-creating employees, now overloaded, burn out and leave the company. The human resource function can play an important role by being the custodian of the employee's interest; in many companies this would be perceived as a full turnaround from the current position. The human resource people in most companies are not looked upon as a service provider to staff.

Many companies, in search of more cost-effective ways of operating, have adopted new concepts such as "empowerment," "total quality management," "team work," and "restructuring," to name a few. Often companies decide to follow a mix of these approaches, which can lead to massive uncertainty and perceived lack of direction and leadership. The human resource function has a role to play in making sure that there is clarity and understanding about what is going on in the company.

■ ***Host governments will continue to become more demanding.*** Host governments have come a long way since the first well was drilled in their country by (typically) a foreign oil company. Initially, most governments were in a weak position due to a need for export revenues and a total lack of experience and trained staff to explore for, produce, and refine hydrocarbons. In the mean time, this situation has changed in a considerable way. The skill base available in these countries has grown to a well-qualified level in many countries. For instance, many countries send selected students out for top university training in the United Kingdom and United States. As a result, these countries insist that the expatriate headcount be reduced in order to bring in their own people.

The human resource function can be instrumental in training expatriate staff to deal with host government issues, to profile jobs in terms of requirements, to identify the right employee to take on a post, and many other functions. This effort requires a massive administrative system and a strong interface with the line function. Another important role for the human resource function is to identify and to train local, high-potential staff. In doing so, the company can develop strong relationships within the host country and with its government. In many cases, these local people will be running the company 20 years later.

■ ***Competition for high-caliber staff is increasing.*** The need for outstanding technical, managerial and commercial skills is stronger than ever before. Obtaining and retaining high caliber staff becomes increasingly important. In its competition with other industries for top-quality individuals, the oil industry is probably worse placed than it was some 20 years ago:

- The oil industry suffers from an unpopular image. Many graduates perceive oil companies as bureaucratic, environmentally unsound, and nonentrepreneurial. The human resource function can influence this image, which is certainly not precise. A range of options can be used to convince graduates to join, including summer programs, presence on university campuses, brochures, and involvement of alumni in recruitment campaigns.
- Attractive alternatives exist outside the sector for graduates, especially when an oil company insists on international mobility. Dual-career couples often prefer to stay in one place, and the salary loss of the spouse is often not compensated for by an expatriation allowance. In addition, the perceived quality of life associated with

expatriation has lost many of its attractions. People who are attracted by foreign countries now can get around through relatively cheap international travel. Staff mobility is a big problem for the oil and gas industry. The need to work wherever oil happens to be found will not go away. Future locations may be even tougher than the ones of today, with new areas such as the former Soviet states and China opening up for the oil companies. The human resource people can be of great use by assessing the scale of the problem, identifying the work conditions that will have to be offered in order to attract staff, and rethinking the need for expatriates within the line functions.

- Oil companies were able to offer attractive careers and salaries to their staff up to the early 1980s, the golden years caused by the oil shocks. There were truly unlimited opportunities, and the limit in one's career was often set by his or her own capabilities and input. Today's situation starkly contrasts with that period and no longer offers unlimited opportunities. Remuneration in the oil industry, although still good, has lost part of its appeal in some companies.

Although much of the above applies equally to the majors, independents, and state-owned companies, the latter have quite a different human resource situation, partly due to their values and pasts. Typical differences include (political) criteria for top management appointments, promotion criteria and timing, remuneration politics, and geographic focus. Specifically for these state-owned companies, a range of additional drivers of change can be distinguished:

1. Most state-owned companies have a strong domestic bias and, in comparison to the non state-owned companies, only limited experience in foreign operations. Figure 12–11 shows how these state owned companies are less advanced in their globalization of human resources.

   However, some of these companies, for example, CPC, OMV, Petrobras, and Statou are slowly but surely stepping into the international arena for a mixture of reasons:

   - The government has changed the mission of the company, and foreign activity has become a real possibility.

● Generous cash flows generated in a domestic monopoly cannot be profitably invested domestically, and foreign projects are believed to offer sufficient profit potential.

The related shift required in human resource capabilities is quite enormous and typically totally underestimated. To put it boldly, the human resource function has to change from a civil-service type of personnel department to a strategically involved function that is highly integrated with the line functions and company-planning procedures.

2. The current wave of privatizations across the world will affect some of these companies. The pressure to meet internationally competitive performance levels will be instantaneous and have a direct impact on the human resource side.

		CURRENT SITUATION			
	COMPANY	Phase	Phase	Phase	Phase
NON STATE OWNED	Amoco	●	●	◒	○
	Arco	●	●	○	○
	BP	●	◒	●	◒
	Chevron	●	●	◒	○
	Conoco	●	●	◒	○
	Elf	●	●	●	◒
	Mobil	●	●	◒	○
	Phillips	●	●	◒	○
	R.D. Shell	●	●	●	●
STATE OWNED	Aramco	●	○	○	○
	Pemex	●	○	○	○
	PDVSA	●	○	○	○
	Statoil	●	●	◒	○
	Petrobras	●	◒	◒	○

● *Fully Applicable*
◒ *Applicable On A Small Scale*
○ *Not Or Hardly Applicable*

Fig. 12–11 Stages for HRD Globalizations.

ONE SIZE DOES NOT FIT ALL

hen trying to improve a company's human resource development processes, a number of pitfalls should be avoided. It must be realized that there is no such thing as one universally best model. Each company has its own model reflecting its history, culture, and geographical presence.

A good example is provided by Royal Dutch, now part of the Shell Group. Royal Dutch hit its first oil in Borneo, far away from its Dutch headquarters and in a time without modern communication. Royal Dutch had to adopt a decentralized operational model, that allowed important decisions to be taken in the field. This model explains Royal Dutch's early awareness that high-quality staff able to operate independently was essential to its success. These people had to be identified, attracted, developed, transferred, and retained by a centralized human resource function. How else could the company get the right people to the right place at the right moment? Royal Dutch, by necessity, took a strategic view on human resources from the start. To conclude from this example that this model is the one to adopt would be a big mistake. There are no guarantees that it will work when introduced in another company's culture, and it is expensive.

Further, a range of lessons can be drawn from the painful mistakes made by companies that aimed to improve their human resource processes:

- Do not expect short-term payoff or underestimate the high cost associated with introducing sophisticated human resource processes.
- Do not leave the job to HR professionals, but instead use a balanced mix of HR professionals and line staff with operational experience.
- Do not start implementation without sufficient preparation.
- Do not underestimate the complications of cross-functional process changes.
- Select top quality people to get process changes in place.
- Ensure top management support and make sure that it is noticed by the entire company.

- Avoid incremental improvements instead of structural changes.

If you respect these rules, you will be well on the way to building the high-performance capabilities your company will need to succeed.

CHAPTER 13

THE EXTENDED ENTERPRISE

Since its earliest days, the oil industry has seen strategic alliances as a means to raise capital and diversify risk. These early alliances were often informal, relying on a handshake and mutual trust. Later, they evolved into more formal joint ventures and joint operating agreements, usually focused on a single field, refinery, or pipeline.

FROM JOINT VENTURES TO STRATEGIC ALLIANCES

oday the concept of strategic alliance must be defined in far broader terms. Take for example, Caltex, one of the oldest joint ventures in the energy business. Founded 57 years ago by Chevron and Texaco, the Caltex flag now flies in over 50 countries. While much of this growth came from expansion of its own operations Caltex has also used key strategic alliances to enter or expand its activities in key markets, for example:

- In the 1950s, Caltex allied with the first Koa Oil Company, then Nippon Oil to enter the Japanese market.
- In the 1960s, Caltex entered Korea through an alliance with Lucky (now Lucky Goldstar)
- Currently Caltex is expanding its operations in Thailand through an alliance with PTT and has also announced an alliance with the government of Oman to pursue joint opportunities in Asia.

Other examples include the strategic alliance of BP and Statoil to develop international gas markets, the "4M+S" alliance (Mitsui, Mitsubishi, McDermott, Marathon, and Shell) to develop part of Sakhalin Island and the Trans-Asian Pipeline alliance (ENI/SNAM, Gaz de France, and Total) to develop a natural gas pipeline network in Southeast Asia. While these are all oil company examples, in fact, strategic alliances have been widely used across a variety of industries for various purposes. Selected examples are shown in Figure 13–1.

The characteristics of successful alliances can be described along several dimensions (see Fig. 13–2):

- Scope: Broad, long-term strategic goals.
- Risk: Extensive, long-term sharing of risk.
- Leadership: Shared decision making through clearly defined processing.

DRIVERS OF COMPLEMENTARY	INDUSTRY	ALLIANCE PARTNERS	MAJOR OBJECTIVES FOR ALLIANCE PARTNERS
ACCESS TECHNOLOGY	Commercial airplane design	Fuji/Kawasaki Mitsubishi — Boeing	Access capital/share risks in new jet development
	Boiling water reactors	Hitachi/General Electric Toshiba	Share risks in second-generation BWR development
ACCESS GEOGRAPHICAL MARKETS	Photocopies	Fuji — Xerox	Access Fuji's technology in minicopiers/access European markets
	Semiconductors	Motorola — Toshiba	Access Motorola's semiconductor technology/access Japanese market
ACCESS NEW MARKET SEGMENTS	Oilfield pumping services	Dowell — Schlumberger	Enter key segment to complete array of oilfield services
	U.S./European airlines	United — British Airways	Access United's U.S. network and reservation system/share facilities
SHARE RISKS	Exploration drilling	Exxon — R.D. Shell	U.K. North Sea exploration and exploitation drilling (NAM)
	Airbus Industrie	Aerospatial MBB BAC CASA	French/West German/British and Spanish consortium
BUILD MANAGEMENT SKILLS	Automotive	General Motors — Toyota	Learn to build cars in a U.S. environment (NUMMI plant)
	Computers	Bull — NEC	Access European markets for computer/develop design skills
ACCESS CAPITAL	Financial Services	Goldman Sachs — Sumitomo	Develop skills in U.S. financial/securities markets
	E&P/LNG	Petroleos de Venezuela — Exxon/Shell Mitsubishi	Gain access to promising Venezuelan reserves/production

Fig. 13–1 Business Leverage from Strategic Alliance.

Comparative Dimensions	Mergers and Acquisitions: Inflexible and Difficult to Successfully Implement	Traditional Joint Ventures: Project-based, limited in Breadth and Scope	Strategic Alliances: Business- or Strategy-based, Broad Breadth and Scope, Adjustable
Time	• No turning back once executed	• Limited time period: clear beginning and end; tactical in essence	• Multibased with broad, long-term, strategic goals
Risk	• Buyer takes all the risks—acquires good and bad characteristics	• Risk-sharing confined to project	• Extensive and long-term risk sharing
Organization	• Buyer usually must implement several difficult restructuring moves (asset sales, layoffs)	• No new formal organization— or skeletal at best	• New formal organization required
Leadership	• New organization, developed during restructuring—often results in clumsy and awkward operations	• Responsibilities clearly defined, leadership focused	• Leadership shared; decisions made through clearly defined processes
Deal Structure	• Complex transactions required for equity and debt financing—may also lead to legal complications (shareholders' lawsuits, etc.)	• Little or no equity/debt financing required	• Little or no equity/debt financing required • Resource allocation based on up front infusion and existing processes within allies • Constant adjustments possible since rigid financial/legal constructions lacking

Fig. 13-2 Strategic Alliances: More Flexible and Powerful than Mergers, Acquisitions or Joint Ventures.

FROM STRATEGIC ALLIANCES TO RELATIONSHIP ENTERPRISES

Even as strategic alliances proliferate—we estimate they have been growing 30% annually over the last four years—an even broader form of strategic alliances, the "relationship enterprise," a term coined by Booz·Allen Vice Chairman Cyrus Freidhiem, is emerging. The relationship enterprise is made up of a network of strategic alliances, linking large (and medium-sized) independent companies. Each has its own agenda and objectives, but they work together to achieve certain strategic goals.

One such example is the recent agreement of six major telecommunications companies to build a global fiber optic network. Another may result from discussions underway among Boeing, Airbus, and several Japanese companies to jointly develop the next generation of jumbo jets.

In oil the opening up of new regions such as Argentina, China, the former USSR and Venezuela has spawned a growing number of alliances which may one day grow into relationship enterprises. The "4M +S" consortium to develop offshore Sakhalin Island is perhaps the most ambitious of these, although a number of other combinations have also emerged, including British Gas/AGIP and Mobil/Exxon.

The continued role of national oil companies with absolute or at least preferential access to large oil and gas reserves, but lacking (in varying degrees) management skills, technology, and (sometimes) capital are natural partners for strategic alliances today. Some like PDVSA and Saudi Aramco have taken the lead in developing overseas alliances. (PDVSA-CITGO-Ruhr Oel and the Star joint venture between Saudi Aramco and Texaco). These alliances may well develop into full-fledged relationship enterprises over the next decade.

This process should evolve over time from single purpose alliances to ones with broader common objectives. Relationship enterprises will tend to develop under the following set of conditions:

- Where size is competitively important such as in industries with major scale advantages, e.g., R&D and capital intensive industries.
- Where the magnitude of the task exceeds the ability of any company to take on the risk. Consider, for example, the development and commercialization of Russia's enormous oil and gas reserves, or a paperless documentation and payment system for the financial world.
- When being global is important to serve customers, but country regulations and laws or customs won't permit local ownership. We saw this when the accounting profession formed relationship enterprises in the 1970s. Airlines and telecommunications have also followed this path.
- Where substantially more capital must be raised to take on a mission and capital markets in other countries are important to the task at hand
- When the national playing field is not level for foreign-owned companies. The auto industry in Japan comes quickly to mind, as do several industries in which national companies receive government incentives/assistance, but foreign companies don't.
- When the potential partner with the right capa-

bilities cannot be acquired for whatever reason

- Where an industry is in critical need of restructuring, but the restructuring won't happen naturally

Relationship enterprises have many advantages. But they also have some disadvantages:

- They are tough to manage. After all, the partners are usually fiercely independent companies.
- They are tough to harness. It is difficult to exercise their full power.
- They could limit competition (Governments, beware).
- It is tough to assure equitable distribution of returns among the partners.
- They require managerial and leadership skills different from the kinds that most corporations have been developing for the past 50 years.

If you are not a part of one, your world could be very difficult. As a group of fundamentally independent entities, the enterprise, by its very nature, is not one that can be easily managed. Today, the prevailing managerial philosophy is built around control. But conventional control orientation won't work in managing a relationship enterprise because the key assets of the enterprise—i.e., other company's—are under someone else's control.

The governing philosophy will have to be based on a recognition of interdependence. The enterprise as a whole must realize that members' needs will sometimes dictate that they go outside the enterprise. Members of the enterprise must be able to deal with shifting alliances and must accept that sharing equitably in the returns can only be considered over a long time frame.

These are real challenges. Partners are going to have to learn to operate in a very different environment. The winners will move quickly to identify and lock up the best partners. They will quickly realize that it is necessary to build strong bridges and highways between organizations to make alliances work.

To succeed in this new world, the first step should be to create a vision of industry. How are the networks likely to evolve? How can that evolution be used for competitive advantage? Capabilities are the fundamental building blocks of these enterprises. Understanding the capabilities that will determine competitive advantage and assuring that those capabilities are developed is critical.

Then comes the hard part—constructing the enterprise. There aren't a large number of successful alliance models. There are six cardinal rules in constructing a relationship enterprise.

- ***Vision is crucial.*** Build the enterprise around the capabilities that will give it important competitive advantages. You need to have a long-term vision for the relationship. It should be a vision that your partners share.
- ***The right partners.*** Choose well for the right reason. Make this a top priority for top management.
- ***Offer the best and the brightest.*** Put your top people into the enterprise. It is the easiest way of telling your partners you think it is important.
- ***Communicate.*** Invest in an infrastructure that makes it easy to communicate and do business together.
- ***Equability.*** Construct the enterprise to serve the interests of all parties as equitably as possible. At all times you should be able to explain what you—and your partners—will be getting out of the enterprise. Companies, like people, tend to act their own self-interest.
- ***You don't have to be boss.*** Construct the enterprise to win. Being king of a losing cause may be far less rewarding than battalion leader in the winning army.

The emergence of these relationship enterprises will pose real challenges both internally and externally. Internally, the challenge will be to manage across the boundaries of two or more corporate structures and cultures. The need to reconcile different information systems, performance measures, global human resources policies, different technologies, and other factors, will pose a continuing set of difficulties.

Externally, governments will find it difficult to influence/regulate these large networks of companies, networks which may not have any formal legal existence in their countries.

Nevertheless we expect that the expansion of strategic alliances and the evolution of relationship enterprises will have a profound effect of the energy markets of the next century.

SECTION IV
THE SPECIAL CASE OF NATIONAL COMPANIES

CHAPTER 14

COMMERCIALIZATION

National oil companies have grown dramatically over the past two decades, both in numbers and in strategic importance (see Chapter 1 for a brief history of their growth). Today they control over 90% of the world's oil and gas resources and are responsible for over 75% of the world's daily production of oil and gas. In fact, of the 50 largest oil companies in the world, 33 are wholly or partially government owned. And their role will most likely increase with time, given their dominant reserve position.

However it can be misleading to view NOCs as a homogeneous group. As illustrated in Figure 14–1, they differ on a number of important dimensions.

	PRIMARILY UPSTREAM	INTEGRATED	PRIMARILY DOWNSTREAM
NET OIL IMPORTERS		ENI Petrobras Petrocanada Elf Total OMV YPF	Repsol CPC Petroperu Neste Oy
NET OIL EXPORTERS	NIOC INOC Pertamina Sonatrach CNPC CEPE Ecopetrol QGPC ADNOC EGPC	Saudi Aramco PDVSA Pemex KPC Statoil Rosneft Lukoil Yukos LNOC	

Fig. 14–1 Characteristics of Selected Nocs.

The national oil companies most likely to be privatized are those in the countries which are net importers of oil and where the oil industry has a relatively modest share of national GNP. Companies in oil-exporting countries where the oil industry is a major contributor to GNP are less likely to be privatized, although we expect some of them will eventually be turned into private or at least semiprivate companies, albeit with continuing government oversight. Indeed debates over privatization of KPC are already a regular topic of debate in Kuwait.

Whatever the eventual ownership outcome, NOCs have clearly entered an unprecedented era of change which will transform the petroleum market as profoundly as the nationalizations which created many of them two decades or more ago. From the late 1960s through the mid-1980s, the trend in most petroleum exporting (and many importing) countries was characterized by increasing state control of petroleum markets. In addition to the establishment or rapid expansion of state-owned enterprises (SDEs), many governments also restricted access of foreign investment capital and imposed varying degrees of price regulation of

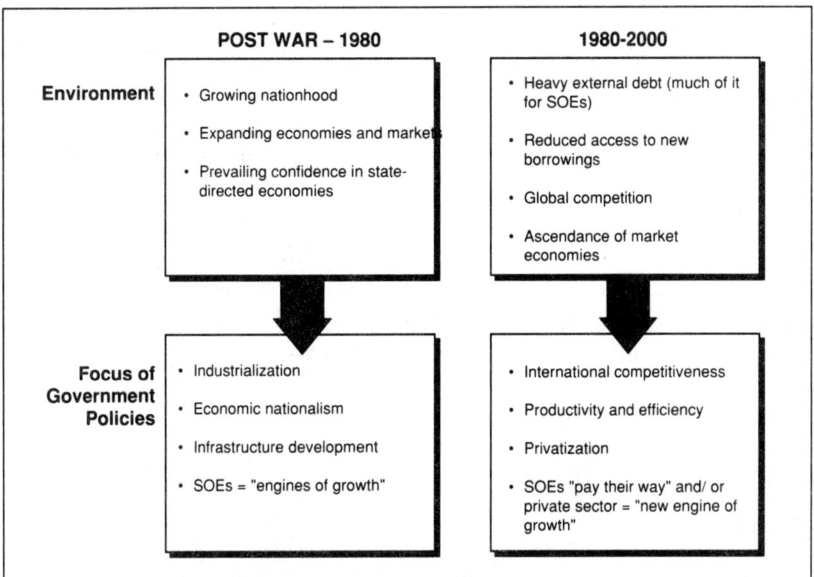

Fig. 14–2 Changing Environment in Developed Countries.

their domestic markets. (In most producing countries, the price was set well below market.)

Since the mid-1980s, however, the trend has been partially reversed. Stimulated by the renewed faith in private ownership, escalating capital requirements, and government shareholders' dissatisfaction with past performance, national oil companies are being restructured and revitalized by a new generation of managers who, in some cases, aim to create globally competitive companies to rival the traditional majors. These changes in part reflect broader trends in many developing countries (see Fig. 14–2).

At the same time, governments throughout the world are liberalizing markets, improving—indeed encouraging foreign investment—deregulating prices and eliminating the formerly sacred idea of government monopoly.

Responding to this external stimulus, NOCs are pursuing a variety of strategies ranging from simply placing an increased emphasis on commercial performance to outright privatization. Collectively, these strategies are changing the very nature of these companies and are laying the groundwork for a global petroleum market that is more competitive and therefore, more efficient. However, the process of transformation has just begun and will take several more years to have its full effect.

Commercialization is the most common strategy being adopted today. Also called corporatization, this is the fundamental process of turning a bureaucracy into a business. Examples abound, from the creation of joint stock companies in Russia to the recent restructuring of Pemex. Governments are insisting that their national oil companies begin to behave much more like their privately owned counterparts. In some cases, this process may be the first step towards privatization; in others, governments remain committed to state ownership.

The commercialization strategy requires insulation of the company from micromanagement by the government and a profound metamorphosis of the corporate culture. Insulation is most effectively achieved by the creation of a holding company ("corporatization") to stand between the government and the operating companies. PDVSA serves as an example (see Fig. 14–3).

The complete cultural metamorphosis involves more than a change of attitudes: it requires major changes in management style, processes, and operations. The most dramatic changes involve the delegation of authority, combined with establishment of performance standards and the introduction of total quality management principles (see Fig. 14–4).

It also often requires rationalization—shedding underperforming assets and/or noncore businesses that have often been dumped into the national oil company as a last resort for otherwise bankrupt entities. The desire of ENI to leave the hotel, textile, and newspaper businesses and the sales of marginal fields by YPF are two examples.

Once a national company has been turned from a bureaucracy into a business and sheds its noncore assets, the next step is to optimize the performance of its remaining core assets. This process requires an alignment of strategy, process and organization, reinforced by tighter performance standards, linked to performance evaluations and compensation. It also implies a strengthened planning process that allows the company to adapt its strategy as the market changes. The discussions in the preceding chapters of achieving success in both the upstream and downstream apply equally to national companies. The Booz·Allen Performance Wheel discussed in previous chapters provides a useful guide to this process (see Fig. 14–5).

The most difficult transformation for many NOC managers is to bring focus to their strategies. Too often the broad objectives of economic development are used to justify any expenditure (no matter how unlikely to be profitable) and any increase in staff (no matter now inefficient). In the future these managers will need to agree on a new vision for the company, to build the capabilities necessary for success, to set quantified objectives to be achieved, and to reinforce those objectives with accountability and a straightforward system of incentives and disincentives.

The acquisition or expansion of downstream or upstream assets has been widely pursued by the national companies of both producing and

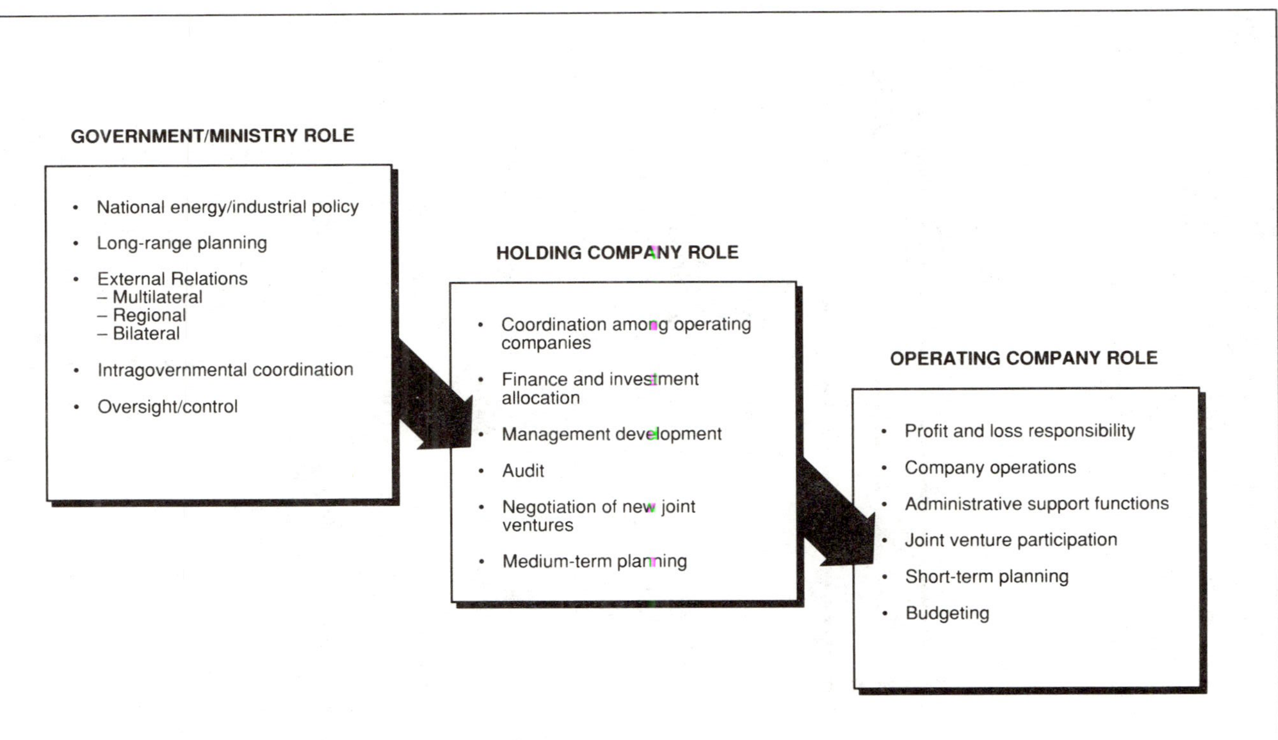

Fig. 14–3 Required Changes in Corporate Culture.

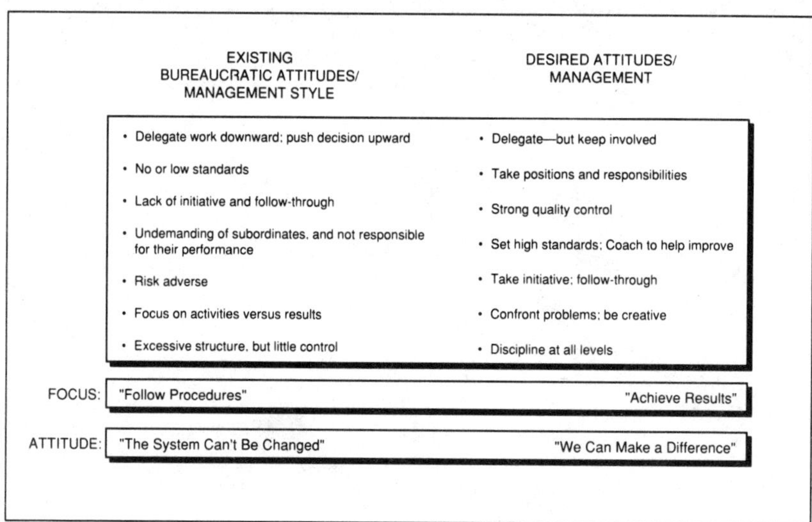

Fig. 14-4 Insulation of Operations From Politics.

consuming nations. The most highly publicized moves have been the downstream acquisitions by Saudi Aramco, Kuwait Petroleum , PDVSA, and the Libyan National Oil Company (through Tamoil). In addition, joint ventures, such as the Pemex-Shell Deer Park Refinery agreement, is also a popular tool.

However, the results of downstream integration by producer companies to date has not been universally financially rewarding and needs to be considered carefully, as discussed in Chapter 7. Integration in gas has taken the form of LNG, methanol, MTBE or petrochemical plants, and/or the construction of long distance pipelines. Here too, integration for integration's sake has not been uniformly profitable.

Globalization is another strategy that has often been linked to integration, as national companies expand outside their home country borders. The most publicized examples of globalization have been the downstream acquisitions mentioned previously by OPEC NOCs, but a number of non-OPEC companies, including Statoil, the Austrian National Oil Company and OMV, have also ventured downstream outside their home countries. Upstream integration has been the driving force behind the globalization of many consumer country national oil companies, including Elf, Total, ENI, Repsol, and CPC. Few of these companies, however, have yet adapted their cultures and human resource strategy to their new multinational/ multicultural identities.

Privatization is the most controversial and dramatic strategy for the rapid and radical transformation of the company. However, only a few

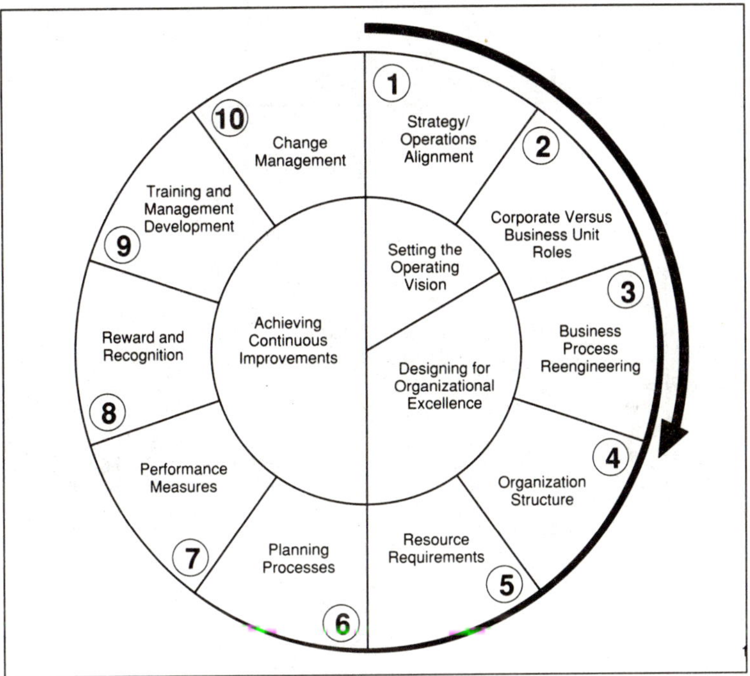

Fig. 14–5 The Booz·Allen Performance Wheel.

national oil companies have been fully privatized. (Examples include BP, British Gas, Enterprise Oil, Veba, Minol, Petrocorp in New Zealand, and Singapore Petroleum Corporation.) And a few have been partially privatized (Elf, Total, Repsol , OMV, and (prospectively) ENI) (see Fig. 14–6).

The net effect of these developments will have profound effects on world oil markets and on the leading actors in those markets. We expect that from the ranks of today's national companies will emerge new major companies. Among the likely candidates are Saudi Aramco, PDVSA, KPC, British Gas, Rosneft, Gazprom, ENI, Elf, Total , Pemex, and Repsol.

Everywhere stakeholders and shareholders will demand higher levels of financial and environmental performance from national, international, and regional companies, who will increasingly find themselves operating in free markets and often in strategic alliances with other companies.

The net effect of these trends may include :

- Increased competition in the world's petroleum markets, which will in turn drive down costs. The growth of competition will benefit consumers and those compa-

OBJECTIVES	CURRENT STATUS			
	Commercialization	Demonopolization	Partial Privatization	Complete Privatization
Complete Privitization			PetroCanada Repsol Total Elf YPF ENI	British Gas BP
Partial Privatization		Rosneft Petronas ONGC	Gazprom Yukos Slavneft Petroleas Del Peru Siberian Far Eastern Surgutneftegaz Norsk Hydro Lukoil	
Demonopolization	INOC Petrobras	Sonatrach NNPC PDVSA ADNOC Pertamina SINOPEC Statoil EGPC Libya NOC PDO QGPC NIOC		
Commercialization	Petrobras Saudi/Aramco KPC PEMEX			

Fig. 14–6 Selected Oil Companies: Privatization Status.

nies that adapt best to the changed circumstances. National companies that fail to adapt will risk facing rising costs and losing of market share as their more flexible colleagues equip themselves for petroleum's second century.

● Competition may also link regional markets more closely . As companies compete, they will seek out the highest returns, effectively arbitraging differences between regional markets. The ability of government ministries to segregate regional markets will be superseded by the profit-seeking interests of companies.

● Strategic alliances and "extended enterprises" linking national oil companies and privately owned companies will proliferate, as the political barriers to such cooperation diminish and commercially motivated partners seek to expand. The Statoil-BP alliance and the Pemex-Shell agreement are but two examples of this new pragmatism. These types of agreements will only work if the two companies pursue commercially congruent objectives.

● Increase the rate of diffusion of information and technology throughout the world petroleum industry. Reduced governmental intervention and subsidization will provide the incentive for an accelerating pace of change and innovation.

● Change the role of governments in many countries from active owners to passive observers and regulators.

Change is never comfortable, but it provides unprecedented opportunities for those companies which recognize the need to adapt and act. Their success—and the failure of others to react—will have profound effects on the petroleum market of the next decade. Let us now turn to a more detailed consideration of the opportunities and challenges of privatization.

CHAPTER 15

THE CASE FOR PRIVATIZATION

Privatization is a major topic these days in the world of oil companies. It is easy to understand why:

- Most of the world's oil and gas assets are in state hands. For example, of the 50 largest petroleum companies, 33 are state owned. These national oil companies own a huge share of world's oil and gas reserves (see Fig. 15–1).

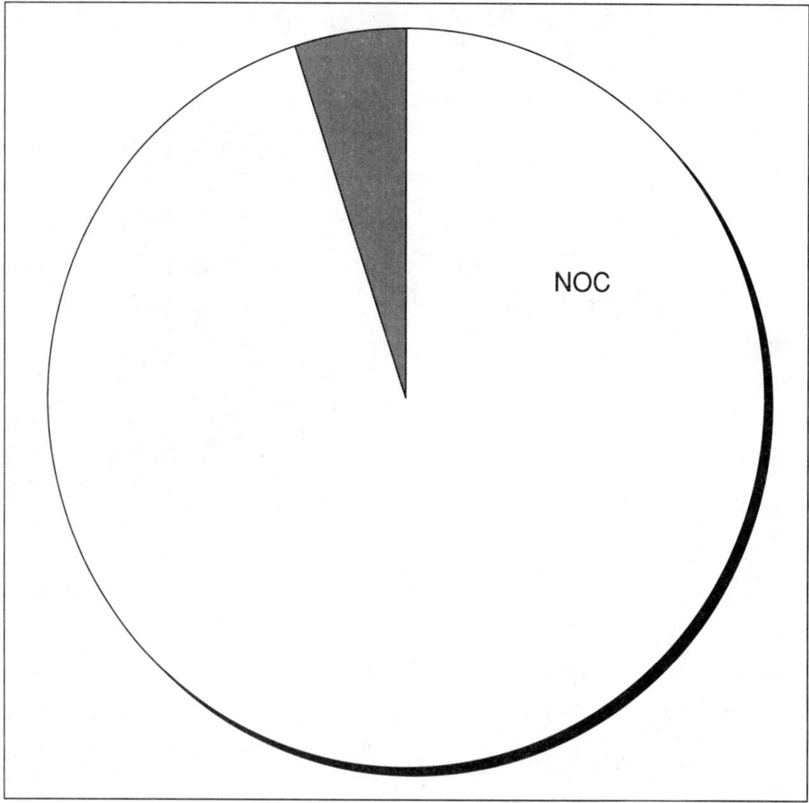

Fig. 15–1 NOC Share of World Oil and Gas Reserves.

- There is thus much to privatize in times where many governments around the world have decided that their countries are better off with their largest industrial companies in private rather than state hands (see Fig. 15–2).
- The list of successful oil and gas privatizations is long—and growing: in Europe, Elf has joined the ranks of BP, British Gas, and Total; Agip and Repsol are close behind. In South America, Argentina (YPF, Gas del Estado) and Chile (ENAP) have led the trend, with Peru and Bolivia not far behind. Elsewhere in the world, the selling to the public of shares in state-owned enterprises (SOEs) may be the prelude to complete privatizations. Recent examples include Malaysia, Thailand, Singapore, and the Philippines.

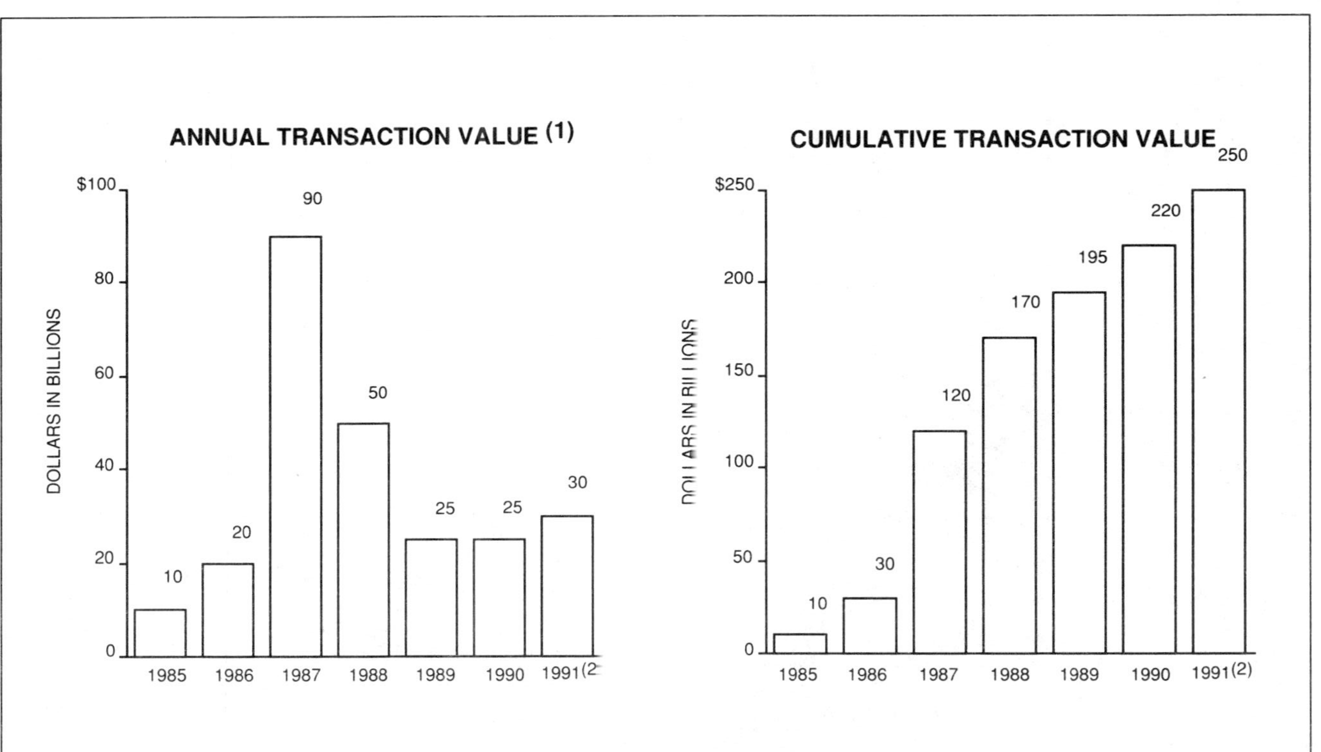

Fig. 15–2 Privatization 1985-1991. Source: Reason Foundation.

Furthermore (and it is interesting to note), in no country has the trend been reversed. No government has "renationalized" an oil and gas company privatized within the last 15 years. This chapter explores why governments are privatizing oil and gas SOEs, then looks at what is involved in a successful privatization. We conclude by looking at a few recent examples in the United Kingdom, Argentina, France, and developing countries.

WHY DO GOVERNMENTS HAVE SOEs?

uring the past decade, government interest in privatizations has been driven primarily by the need to reduce national debt (or meet future capital requirements) and by the superior performance of market-based economies in the face of the collapse of centrally planned systems. There has also been clear evidence of the superior performance of private companies relative to SOEs.

The specific reasons for privatizing major oil and gas enterprises vary country by country. However, without claiming to be exhaustive, the following socioeconomic and political reasons can be listed:

- Improve national productivity
- Reduce national debts
- Get an influx of capital to help balance socioeconomic programs
- Lower future energy costs
- Raise future energy export earnings
- Deintegrate inefficient monopolies
- Increase domestic competition
- Take money-losing enterprises off the state's books, and (eventually) get additional taxes from profitable private companies instead. The lack of inherent profitability of SOEs, most of which lose money with great regularity, makes the issue of "selling the silver" moot. In other words, the long-term profitability (and income tax–generating capability) of the sector is much more important than the actual proceeds of the privatization sale itself. That's true even if it is the sale that makes most headlines.

- Broaden the pool of shareholders (from the state to the public, both within the country and internationally, bring in world class institutional investors).
- Stimulate local capital markets.
- Improve management skills—both within the former SOEs and at the senior levels—leading to the emergence of market-oriented senior executives within the company.
- Meet a political agenda—such as reducing the role of the state.
- Allow the state to focus on infrastructure (social programs, health care, Social Security, retirement and unemployment benefits, pension funds, education, and the like), rather than on running companies, which the private sector can do better.

In summary, many governments believe that, if properly planned and executed, privatization can create globally competitive companies and provide substantial benefits to the country and the company (see Fig. 15–3).

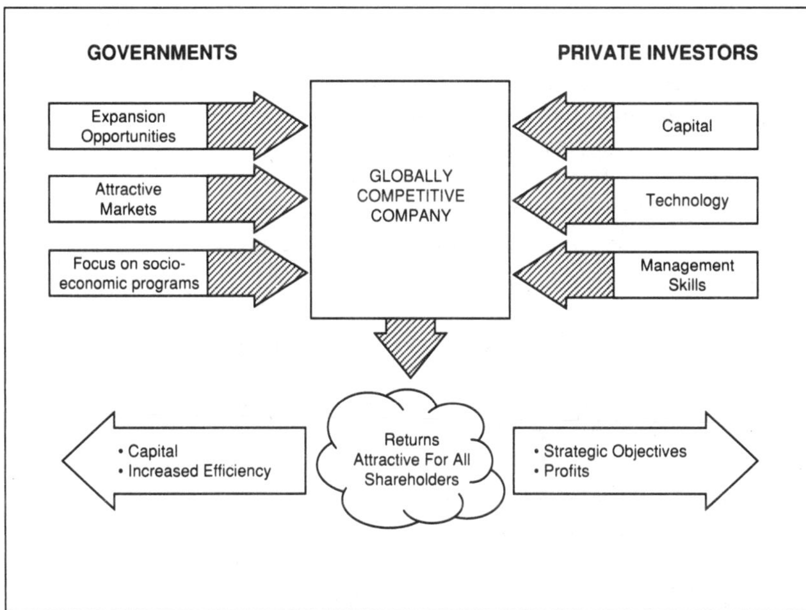

Fig. 15–3 Benefits of Privatization.

PRIVATIZATION—A GENERIC CASE

To understand the complexity and challenges of a large SOE privatization, one needs only to look at the list of actors involved:

- Local government
- The soon-to-be-sold SOE and its senior executives, middle management, and (often unionized) employees
- Financial investors: stock market shareholders, large institutional funds, global financial markets
- Large institutional investors (e.g., oil and gas majors, other energy companies with an aggressive international strategy)
- Local investors (e.g., independents)
- Local financial institutions
- Local stakeholders, beyond SOE employees and management (e.g., provincial governments)
- Regulators (if any exist prior to the privatization process)
- Privatization committees—often formed with government and industry officials to prepare the SOE or sector at large

With all these players involved, each one with its own nationality, background, and agenda, a clear privatization strategy is necessary to channel all institutional actions in the process to ensure a win-win situation for both the local government and investors.

Let's now look at what the government strategy should include.

(LOCAL) GOVERNMENT STRATEGY

The first decision the local government has to take is how far to go. As already seen in the prior chapter on commercialization/corporatization, there are several potential intermediate steps on the path toward privatization. Several government owners of prominent oil companies have set their current objectives on these intermediate steps. For example:

- Commercialization (Pemex in Mexico and Petrobras in Brazil)

- Demonopolization (Petroleos de Venezuela, PDVSA)
- Partial privatization (Repsol in Spain)
- Recapitalization/investments by other, private, major energy companies (YPFB in Bolivia, Bayernwerck in OMV)
- Major (i.e., multibillion dollar) alliances (PDVSA with Exxon/R.D. Shell on the Christobal Colon LNG project, Sinochem with Amoco)

The second critical step is for the local government to develop a clear vision of its oil and gas sector and how it is likely to evolve over the long term. This vision should include three strategic components:

1. ***The Degree of Competition.*** Depending on the degree of competition desired, the vision for the sector can move along a continuum of which the two extremes are:

- Privatized monopolies; the most noteworthy is British Gas in the United Kingdom.
- Break-up of an integrated SOE along several stages of the sector, each one including one or several newly created, publicly traded companies. This model looks at the deintegration of a vertical energy supply chain. Usually, this means splitting the oil, natural gas, or electricity supply chain into production (generation), transmission (transport), and distribution companies.

In the case of the former U.K. electricity monopoly, the Central Electricity Generating Board (CEGB), privatization of the sector led to two generators (not to mention the still-owned nuclear authority), one transmission company, an electricity market exchange to determine prices for the system, and twelve regional distributors.

In Argentina, a similar model for the privatization of the natural gas assets of the former Gas del Estado led to the creation of two long-distance pipelines and eight distribution companies.

2. ***Clear objectives for the desired anticipated benefits of the government strategy.*** For example:

- Obtaining short-term financial gains (to help alleviate temporary government budget deficits, such as in the United Kingdom during the sale of BP, British Gas, or in France during the privatization of Elf).

- Creation of a truly competitive sector, even in sectors often considered natural monopolies (such as the U.K.CEGB/electric sector, or Argentina's Gas del Estado sale of natural gas assets).
- Attraction of international investors. Institutional, as in the example of YPF in Argentina Direct participation by prestigious international energy firms—such as Argentina with Gas del Estado (with British Gas, Enron, Nova, Italgas, and other international gas companies).
- Improvement of overall economic efficiency for the sector. Some SOEs lose hundreds of millions of dollars every year. When this is the case for several SOEs at the same time, most governments feel compelled to take action. Some governments prefer to fix the money-losing SOEs, others elect to sell them "as is" to the private sector, assuming that the private sector will be better equipped to bring them back to profitability. As we will discuss in a moment, that does seem to be the best approach.
- Sometimes, partial or total privatizations can be used to achieve a gradual transition from a centrally planned to a more market oriented economy. This is the case in Russia, with the current, ongoing partial privatization of the world's largest national gas company, Gazprom.

3. ***Envisioning a clearly defined endgame is a must, and should be done with both quantitative and qualitative parameters in mind.*** Among the questions that need to be asked and answered are:

- ***Quantitative.*** What percentage of private sector participation should there be in the former SOE? Percentage of international investment relative to the total proceeds of the privatization? When and how much of the company should the government sell?
- ***Qualitative.*** Extent of demonopolization? Type and number of companies created after the break-up of a vertically integrated SOE? Breadth and scope of business activities of the newly created companies?

Once the government has determined a clear vision and strategy, the vision has to be marketed actively. A clear vision strategy presented to the potential participants in the privatization program can achieve an enormous amount of confidence building among potential investors. And the willingness of these parties to invest hard-earned profits into the privatized firm is ultimately a major (if not the only) yardstick of success of the operation.

Governments who take the time to explain and promote their privatization programs (1) to potential local and international investors, (2) to the financial community at large (local markets, Wall Street, London, Tokyo, etc.), and (3) to all other stakeholders, both within and outside the country, can achieve a tremendous amount of goodwill among the main actors of the privatization. This is particularly important in the case of governments with no prior track record in this type of operation.

In the United Kingdom (Thatcher's privatization program) and in France (the 1986–88 and 1993–95 programs), much consideration was given to the early privatizations—their success or failure being perceived as having important repercussions on the whole program. This does not mean that governments are not allowed to go through a privatization learning curve. They are, as long as they send clear signals that early mistakes are understood and will be corrected in time for subsequent privatizations.

In summary, there are as many strategies for privatization as governments have reasons to privatize. There is one constant requirement, however: to lay out a vision, strategy, and detailed program and communicate it with clarity and transparency to all the players. If this is done, success depends on the thoroughness and quality of execution.

PRIVATIZATION IMPLEMENTATION—CHALLENGES, PITFALLS, AND OPPORTUNITIES

The vision and strategy for a successful privatization revolve around the government; for the most part, privatization implementations are centered around the targeted companies, their executives, and staffs.

Implementation is like a successful 100 meter dash: start aggressively (preparation of the SOE or putting the house in order prior to the sale), concentrate on a strong midrun effort (planning of both the specifics of the financial transaction and the ongoing evolution of the sector in which the newly privatized firm will evolve), and throw everything into the final stretch, (keeping in mind that economic life should go on profitably after that day). If the objective is a sub-10 second time, all three phases of the race have to be performed in winning fashion.

An insufficiently prepared company (not restructured, and/or sold at too high a price) will discourage investors; indeed, the initial offer may even have to be withdrawn ("The government is ripping off small shareholders, widows, and orphans," will be the cry). A poorly planned financial transaction will lead to massively oversubscribed sales ("The government is leaving money on the table").

Poor sector planning will deflate investors' confidence, especially in the case of large international investors. Finally, on D-day (the day when shares are sold by the government), mistakes of execution can lead to all sort of charges (even if unjustified/unsubstantiated) against government and company officials—favoritism, insider deals, incompetence, etc.

The keys to successful privatization include:

1. ***Preparation of SOEs.*** "Putting the house in order prior to the sale." Many SOEs targeted for privatization are in poor financial and operational health. When this is the case, most governments will launch a thorough restructuring program to bring the SOE's operational profitability more in line with private sector benchmarks. Such restructuring involves a series of detailed programs, including development of a high-performance organization, comprehensive reengineering of the main operational, management, and business processes throughout the company, as well as more specific actions, such as revamping the firm's information systems. Booz·Allen describes how all these restructuring steps fit together in the High-Performance Organization Wheel. Examples of such restructuring prior to privatization in the oil and gas world include the ongoing reorganization of Italy's ENI Group and Repsol in Spain and at Yacimientos Petroliferos Fiscales (YPF–Argentina).

On the other hand, many governments seem to overestimate the value of such restructuring prior to the actual privatization. The successful—indeed spectacular—transformation of British Airways and the suburban railways of Buenos Aires argues for letting the private sector, and not government, restructure the SOE.

The necessary metamorphosis of an SOE into a successful private firm thriving amid intense competition involves much more than restructuring; it also requires significant changes in attitudes and management styles. That too is easier to achieve with an entirely new regime for the company than within the existing state boundaries and bureaucracy. To be fair toward SOEs and government civil servants, there are national companies that perform admirably and need relatively little preparation to allow a successful privatization effort to take place. BP (around the time of privatization), British Gas, Total (with a very gradual sell-off of the French state's equity stakes), and more recently, Elf Aquitaine, all support this point.

2. ***Planning the Specific Transactions and Preparing the Sector at Large.*** The planning of a large privatization can be a delicate financial affair. In the case of oil and gas firms, they are often the largest companies in their countries, by revenues or (potential) market capitalization. For example, among the companies recently privatized or in the process of being privatized, Elf Aquitaine (France), YPF (Argentina), and Petroperu (Peru) are the largest firms in their respective countries. During their privatization process, various ministries (energy, treasury, finance and economics, infrastructure/industry, planning), local financial institutions and international banks (from Wall Street, the city, and often the World Bank) have to work hand-in-hand to ensure smooth implementation.

As discussed in the previous section, the preparation of the sector-at-large requires due diligence including detailed implementation plans outlining:

- Required institutional changes (such as role of government/energy ministry after the sale and new legal and regulatory frameworks)
- New roles for government (at the federal or central level) and provincial/regional authorities (balance of power, interfaces with regional authorities, and infrastructures in the case of transportation and distribution networks)
- Creation of new regulatory bodies. The creation of new networks of public oversight can be of paramount importance, and should take place in different fashions, depending on the type of energy sector involved.

  When so-called natural monopolies such as natural gas and/or electricity transmission and distribution networks are being privatized, there is a strong case for creation of a new regulatory body (such as Offer and Ofgas in the United Kingdom). Such an agency can maintain an arm's length balance between the companies and their customers, achieve a delicate equilibrium between the interests of investors, company management, and customers, all the while ensuring that the overall networks are operated safely and without curtailments.

A different approach is called for when an oil and gas SOE is privatized in a producing country. In this case, attracting international exploration and production capital and major oil companies is at least as important as the privatization process itself. The government therefore needs to pay careful attention to normative and operational functions within the SOE. Often the SOE is responsible for establishing negotiating guidelines with foreign E&P companies (such as the policies governing service/production-sharing contracts, royalties on production, fiscal regimes, profit repatriation rules, and the like). The SOE is also often responsible for negotiating the actual E&P contracts with interested parties, as well as for the operations within a large share of the local hydrocarbon bearing properties and exploration acreage. Clearly, prior to the privatization process, the government has to address these issues to the satisfaction of would-be foreign E&P operators. This may involve the creation of an E&P negotiating body at arm's length from the SOE and preferably separate from the Ministry of Energy as well.

- Other factors not to be overlooked during this due diligence phase include:

■ ***Preparation of Local Public Opinion.*** This is necessary to achieve the desired political changes.

■ ***Financial Reform and New Fiscal Regimes.*** While often profitable, SOEs are not always run as efficiently as they should be.

■ ***Dealing with Assets of the SOE that Prove Not to be Easy to Privatize.*** There are solutions. They can be abandoned or written off, grouped with another SOE, or as in the case of the United Kingdom's nuclear plants, put into a new SOE.

3. ***Management of Multiple Transactions Involved During the Actual Privatization.*** The actual privatization (and the few months leading to it) involves myriad steps, each best performed by a group of specialists whose participation is planned well ahead of time. These specialists must tend to:

- Marketing to local and international investors
- Preparing public tenders

- Detailed financial issues
- Detailed legal issues
- Detailed technical issues
- Completion of supporting consulting studies
- Preparation of actual bidding process
- Calendar management
- Week by week planning of the transition period from public to private ownership
- Planning for contingencies: how to keep flexibility for necessary last-minute course changes and other unforeseen obstacles

It is also important—despite the huge pressure building up to the day of privatization—to keep a vigilant eye on the industry or sector and its evolution after D-day. Therefore, some more detailed implementation work with newly established regulators and government agencies has to be undertaken ahead of time to ensure that:

- There is ongoing monitoring of the industry from a safety, technical compliance, and fair competitive standpoint. This monitoring should include contingency plans to fight against potential future oligopolistic positions (e.g., in power generation) and against natural but privatized monopolies abusing their position to overcharge/underservice their customers.
- The newly created regulatory bodies (upon which will fall the responsibility to monitor competitive forces, safety, technical maintenance required, new investments, customer service/complaints, and reporting) should be adequately organized, staffed, and prepared within the months prior to privatization.

ILLUSTRATIVE EXAMPLES

When the Thatcher government decided in the 1980s to embark on a sweeping privatization program, it had a clear ideological and political agenda. One of the advantages of a clear ideological and political agenda is that each privatization fits within a well-organized program, with clear financial synergies. In addition, lessons learned during one privatization can be applied to another. Privatizations would be a key element of a profound transformation of the U.K. economy. Groups of very different natures, from airlines and railroads to major oil and all types of utilities (telecom, natural gas, electricity, water), were included in the program. The purpose of this section is not a wholesale review of the U.K. privatization program—whole libraries have been written on this topic—but a quick demonstration on how the privatization of British Petroleum and British Gas illustrates the implementation guidelines described earlier in the chapter.

The privatization of British Petroleum represented a milestone in the history of major oil companies: One of the original "Seven Sisters," with revenues above $50 billion, became a private company. It was arguably the largest transformation of such a kind up to then.

It is fair to say that BP up to that point had essentially been run much like a publicly traded group. As a result, there was no need to prepare the U.K. exploration and production, downstream, and chemical sectors: they were already fully exposed to intense domestic and international competition.

As for BP itself, one of the main internal efforts had be to digest the recent acquisitions of Britoil and Sohio, but there was little restructuring needed, only minor "window dressing" required prior to the privatization.

As a result, the company could be auctioned off pretty much as it was, and all the U.K. government efforts focused on the ability of the city to "digest" such a formidable public offering. The sale of BP shares in a relatively soft market was a great success, but one key investor (Kuwait) was in a way penalized for having helped—being pushed into reselling a significant equity share at relatively low prices due to political sensitivities within the United Kingdom.

Overall, BP's privatization was a remarkable success, and the ups, downs, and ups of the company after privatization have shown that it truly behaves like an international major, operating in a very competitive environment, with corresponding penalties for low performance and rewards for success.

British Gas was quite a different story than to BP, in many ways much closer to other giant U.K. utilities such as British Telecom and the Central Electricity Generating Board. For all these utilities, one key issue was whether to privatize them whole or split them along production (natural gas E&P, power generation), long distance transmission, and local/regional distribution lines. In such companies, there is always a fear that an integrated, privatized quasi-monopoly might use its transmission network—the backbone of the system, into which everything supplies and feeds—as a leverage to capture above-normal returns all along the supply chain.

In the case of British Gas, there were two additional concerns that proved to be dominant in the government officials' minds:

1. International competition is growing stronger every year. In particular, in Europe, the need to rely on many nonindigenous sources for natural gas supplies creates extraordinary demands on investments: While Groningen gas is cheap, U.K. and Norwegian North Sea gas suppliers have required huge investments. Algerian gas flows either through expensive LNG supply chains or through pipelines either in need of expansion (Transmed to Italy) or in need of construction (new link to Spain). Russia is the last main source of natural gas for Europe, and is even more expensive. And future Russian/Siberian fields are going to cost even more to develop, since new multibillion export pipelines must be built in very unstable geopolitical environments.

This supply constraint pushes most European governments to promote and support national champions. Could a split-up British Gas have been able to take on SNAM/Italgas, Gaz de France, Ruhrgas, and its formidable shareholders Royal Dutch Shell, Exxon, BP, Mobil, and all? For the British government, the answer was clearly no. (It is interesting to note that, more recently, the Spanish government made the same assessment and in the early 1990s regrouped the various Spanish gas distribution and transmission companies into one integrated company, Enagas, under the umbrella of the giant oil and gas SOE Repsol.)

2. At the (relatively) early stages of the sweeping privatization program, single integrated companies like British Gas were much simpler to value and offer to the public than deintegrated groups of newly created entities. This meant less risk of over/underpricing, with the corresponding adverse political consequences. And so British Gas was privatized as single, giant entity, and it subsequently proved to be more than able to hold its own against competitors such as SNAM/Italgas, Gaz de France, Ruhrgas, and even Enron.

And what about concerns about dominant position abuses after privatization? In the case of British Gas, all regulatory bodies and mechanisms set up by the British government, Ofgas, the Monopoly and Mergers Commission (MMC), and Board of Trade functioned actively and effectively (if sometimes acrimoniously) with British Gas to balance company shareholders' and customers' interests.

In France, for the second time around after a first wave of privatization during 1986–88, the government advanced new privatizations to disengage the state in a massive way from the operation of companies as diverse as Elf and Total, Rhône Poulenc (chemicals), Péchiney (aluminum), UAP (insurance), BNP (banking), Bull (computers), Air France (airline) and Renault (motor vehicles). Of major oil companies in France, one can note a few differences in what was done in the United Kingdom Specifically:

- The state plans to retain shareholdings in both Elf (21%) and Total (5%).
- The state retains representation on the board of both companies, with veto power for decisions involving 10% of assets or more.
- There is key cross shareholding with other leading French industrial groups.

It is also worth noting that the two other key French energy companies—Electricité de France (EdF) and Gaz de France (GdF)—are conspicuously absent from the privatization program, EdF and GdF employees still being bona fide civil servants. On the equipment vendor side, GEC-Alsthom, which manufactures a variety of fossil fuel power equipment (steam and gas turbines, boilers, controls, etc.), is in private hands and may take control of nuclear power equipment manufacturer Framatome in the near future.

Across the Atlantic, Peru and Argentina offer further instructive examples of privatization. The privatization of Petroperu is interesting because of the careful attention paid to several privatization options. Specifically, Booz·Allen assisted the Peruvian government examine potential groupings of the assets to be privatized were examined, along with the consequences they would have on competition. Also, we examined how attractive those assets would be to potential investors (see Fig. 15–4).

What the Peruvian oil company example illustrates is the variety of privatization alternatives (see Fig. 15–5). There are no cookbook approaches. Each government must analyze and determine the method most appropriate for the situation at hand (see Fig. 15–6). This is particularly true for developing countries where local stock markets do not have enough capitalization to absorb the floating of shares of large SOEs. The need to attract foreign investors into the country (direct asset purchase) or convince foreign institutions to participate in the equity offerings dictates that the effects of each privatization alternative be thoroughly assessed on its potential impact on both domestic and international investors.

The Argentinian privatization program is one of the most ambitious to date. Its comprehensiveness, thorough planning, and successful implementation has led to enormous economic benefits to the country, both in terms of financial success for the local treasury and shareholders, as well as in terms of attracting prominent international companies and institutional investors.

OPTION I	OPTION II	OPTION III	OPTION IV	OPTION V
The Petroperu Plan	**Fragmentation Into Basic Business Units, With Strong Integration Restrictions**	**Fragmentation Into Large Business Units, With Few Integration Restrictions**	**Downsizing Of Petroperu**	**Regional Integrated Companies**
• Petroperu remains as an integrated oil company and is privatized a such • Company is restructured before privatization, including: – Close down of unprofitable units – Subcontracting of services – Services for joint venture partners with capital technology – Reduction of costs	• Petroperu is divided into basic business units • All units are put up for sale immediately • Strong restrictions are imposed to avoid reintegration – E&P areas to be fragmented in several fields – No one is allowed to be in more than one stage of the supply chain – i.e., one company cannot buy an E&P field and a refinery	• Petroperu is divided along the lines of Option II except for the Talara and Selva Norte fields (which remain as simple E&P areas) • All units are put up for sale immediately • Only minor restrictions are applied to prevent dominance of refining and terminals to a single company • Otherwise, reintegration is allowed if buyers desire to buy assets across the supply chain without jeopardizing competition	• Petroperu remains an integrated oil company • Selected core assets are sold to allow free competition to take place • The company is restructured before privatization—with significant downsizing	• Petroperu is divided into large, regional blocks to attract major investors • Restrictions are set to avoid monopolies • Vertical reintegration is allowed, along regional lines

Fig. 15–4 Peru: Privatization Option.

Option	Description
Public Share Offering	• As proposed by Petroperu • First stage of YPF (Argentina) model
Private Sale	• Sale to a single owner/consortium • Gas del Estado (Argentina) model • Enables prequalification—attractive to technical operators
Auction of Block of Shares	• Sale of blocks of shares to several investors—for example, the Brazilian model • Some ownership fragmentation may be achieved • Does not necessarily attract technical operators
Private Sale Followed By Public Share Offering	• Initial sale of control through a private sale arrangement • Subsequent public share offering
Leasing/Processing Agreements	• For less attractive/difficult to sell assets (e.g., refineries) • Could allow the disposal of Conchán without having to resort to a very low price which would be politically unpopular
Shutdown/ Liquidation	• For nonprofitable, nonstrategic assets • If none of the above alternatives worked

Fig. 15–5 Privatization Sales Options.

The privatizations there were even larger, relative to the size of the local economy, than the wave of U.K. privatizations mentioned earlier. Under the leadership of President Carlos Menem and his Minister of Economy and Finance Domingo Cavallo, the following companies were chosen for privatization:

- The national airline, Aerolineas Argentinas (sold to Iberia of Spain)
- Entel, the telephone company (sold to a consortium led by Telefonica of Spain)
- The main freight railroad carrying wheat and other commodities from the Cordoba and Buenos Aires provinces to the port of Bahia Blanca (sold to a consortium led by the RENFE, the Spanish railroads SOE)
- The main port authorities of the country
- The water purification and distribution system of the Buenos Aires province
- Suburban railroads around the greater Buenos Aires metropolitan area
- Other main railroads in the country
- Major toll roads
- Several banks

Virtually all the assets of the energy sector that were not already in the hand of domestic and foreign private operators were also privatized.

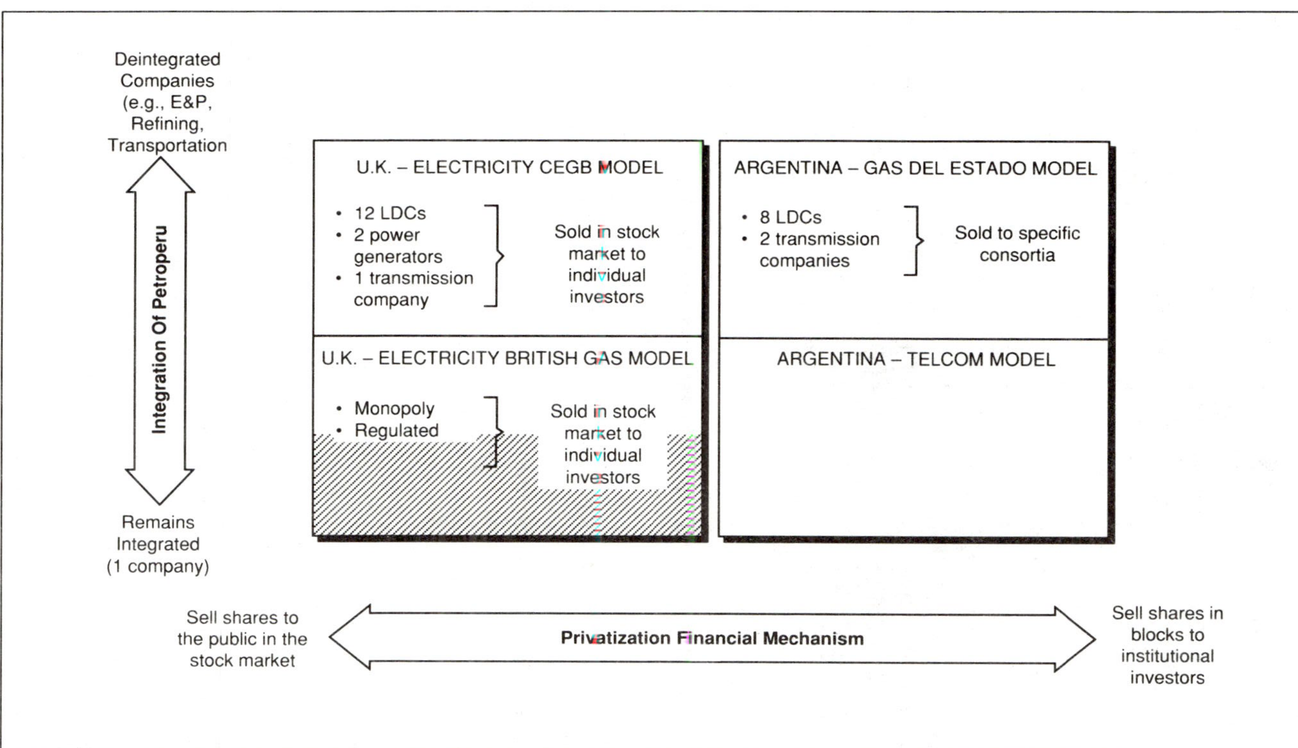

Fig. 15–6 Privatization Alternatives—Examples.

- Yacimientos Petroliferos Fiscales (YPF)
- Gas del Estado (natural gas transmission and distribution assets)
- Several electricity distribution networks, including the assets of the old Segba, which distributed electricity in greater Buenos Aires. It was split into two new companies, Edenor and Edesur

The YPF privatization process went through a milestone in 1993, when an ambitious initial public offering was successfully floated on the Buenos Aires, New York, London, and other international stock exchanges was significantly oversubscribed. This very successful financial transaction capped over five years of intensive efforts, which started with the Houston Plan in the late 1980s to open up exploration and production acreage to international majors. These preprivatization efforts continued with a large sale of marginal E&P and downstream assets owned by YPF, to local independents and international energy companies. (Note that most of these marginal assets are now performing very well for their new owners.)

The objectives behind the privatization moves were twofold:

1. Increase the E&P and oil downstream private sector participation prior to the YPF privatization. Today Agip, Amoco, Braspetro (Petrobras), British Gas, Chevron, Deminex, Exxon, Occidental, Repsol, Royal Dutch Shell, Texaco, and Total—among others—are leading international participants in the sector and competitors of YPF, alongside the main local independents such as Astra, Bridas, Perez Campanc, Pluspetrol, Techint, and Techpetrol.

2. Shrink the size of YPF down to the level at which it could be both a manageable, profitable organization, and one which would not exert quasi-monopolistic power in any segment of the oil and gas supply chain.

In its journey toward complete privatization, the Menem administration developed a vision for the sector, assigned clear financial objectives for the whole process, and laid out a well defined endgame for YPF. This carefully executed multiyear plan, supported by strong marketing efforts targeting international investors, allowed the privatization of YPF and the increased openness of the E&P and downstream oil sectors to be implemented flawlessly, despite the many difficulties associated with the privatization of the largest company of the country.

MAJOR ELEMENTS OF PRIVATIZATION

- Vertical deintegration
- Five-year price caps
- Transmission cannot own gas
- Competition
  - Multiple companies
  - Bypass
  - Third parties (brokers)
- Investment required
- Transition period of one to two years
  - Limits on bypass
  - Contracts given

Fig. 15–7 Gas Del Estado Privatization.

In Gas del Estado a team led by Messrs. José Estenssoro (new chairman and CEO of YPF), Patricio Perkins, and Raul Garcia prepared and implemented a leading-edge privatization for a gas transmission and distribution company.

The Gas del Estado privatization has led to the deintegration of the former SOE and the creation of two gas transmission companies and eight distribution companies. The two transmission companies split the country's pipeline network along a north-south divide, so that each one owns part of the network connecting the rich Neuquen fields with Buenos Aires (west-east), as well as one of the two older trunk lines that connect northern and southern provinces with the capital (see Fig. 15–7).

The major elements of this natural gas privatization were:

1. Vertical deintegration
2. Five-year price caps
3. Transmission companies cannot own gas (transport function only)
4. Competition is encouraged by:

    - Multiple transportation companies (transcos) and distribution companies (discos)
    - Bypass allowed
    - Third parties (gas clearinghouse; brokers)

5. Investment required by new owners (bypass, looping; extensions etc.)
6. Transition period of one to two years with:

    - Contracts/initial tariffs given
    - Limits on bypass

The recent natural gas privatization in Argentina is unique. It has leapfrogged recent experiments in the United Kingdom and in North America.

In the United Kingdom, the privatization of British Gas, although representing a dramatic change (and improvement) in moving a public bureaucracy into private hands, preserved the vertical integration of the industry. Regulatory powers, rather than following a predetermined framework, have been driven by the personalities of the regulators.

In the United States, the natural gas industry evolution toward more competition has been built upon a private industry base with diverse ownership—however:

- The competition created by open access and

unbundling of service offerings by pipelines has often led to acrimonious and expensive litigation.

- Both the Federal Energy Regulatory Commission (FERC) and the state public utility commissions (with ongoing reregulation and revisiting of issues) exercise a strong regulatory hand.
- For most distribution services, cost-plus pricing mechanisms prevail for residential and commercial customers.

In continental Europe proposals (e.g., from the European Community) exist to restructure the industry, though resistance will delay any dramatic changes for the foreseeable future.

The vote of confidence by international gas companies—one of the key objectives and yardsticks of this privatization—has been overwhelming. The sale of Gas del Estado valued the former SOEs assets at about $4.2 billion, about $2 billion more than anticipated. Prestigious international gas companies such as British Gas, Enron, Italgas, and Nova participated in the sale.

The first major investment bid after the privatization—an international pipeline crossing the Andes from Neuquen to the major urban areas of Santiago and Conception in Chile—attracted British Gas, Tenneco, Enron, Transcanada, Nova, and others, with the British Gas-Tenneco led consortium winning the bid.

As in the case of YPF, careful planning helped win over many investors. In particular, the privatization of Gas del Estado relied on a newly established regulatory agency (Ente Argentino Regulador de Gas, ENARGAS, Ente) for implementation in a country with no history of independent regulatory oversight.

Even within a competitive price cap regime, it was perceived that there was still a need for regulation. It was needed:

- To watch out for captive consumers
- To establish and endorse technical and safety standards
- To ensure competition was not subverted
- To review and reset tariffs after five years

The Ente has been given overall responsibility for enforcing the law, as well as for resolving disputes. The law generally requires the Ente to resolve most disputes within 30 to 60 days and to conduct public hearings. Ente decisions can be appealed first to the minister and then to the courts.

The Argentinean government ("subsecretaria de hydrocarburos") wanted answers to key questions about a good operation of the newly privatized gas network before the sale so that no negative surprises would emerge during or after the privatization (see Fig. 15–8).

The key questions involved what specific tradeoffs would have to be made in Argentina between transparency and efficiency, communication among parties, detailed prescriptions vs. broad policies, regional vs. central control, communication among directors and staff, and public hearings vs. delay/backroom deals.

Using insights gained from global experience, Argentina adopted a unique regulatory model embracing the following goals:

- Regulation should be light handed. Make incentives, competition, and markets work whenever possible.
- Protect captive customers.
- Use public hearings as key component of the overall regulatory process.
- Push as much work to companies as possible.
- Use publicity to stimulate performance.
- Stay lean.

The vision developed for the Argentinean gas industry achieved the required tradeoffs to ensure its success. Thus ENARGAS is at the same time:

- How to ensure efficiency?
- What level of openness?
- What tradeoffs between openness and efficiency?
- How far to push the culture?
- How independent?
- How would a public hearing work
- How to meet the deadlines in the act?
- How to ensure due process?

- What will the Ente really have to do?
- How to ensure safety?
- Should the Ente be proactive or reactive?
- What data to collect and publish?
- What should the organization look like?
- What qualifications for directors
- What qualifications for staff?
- How many staff?
- What mix of contractors versus staff?

Fig. 15–8 Regulatory Issues.

- Independent: five commissioners appointed to fixed terms
- Objective: interests of industry and customers are balanced carefully
- Transparent: public hearings
- Efficient: fewer than 100 employees
- Proactive: in monitoring safety, reliability of service, and new investments

The Ente has responsibility and authority for enforcing law and privatization agreements:

1. Specific responsibilities include:

   - Establish and enforce technical safety standards
   - Approve expansion proposals
   - Handle unfair tariff complaints: discrimination, discounting
   - Perform five-year review of tariffs
   - Ensure investment requirements are fulfilled

2. Procedures

   - Follow due process
   - Public hearings in specific cases
   - Tight deadlines: 30 to 60 days

3. Appeals

   - First, to the minister
   - Then, to the courts

Key policies were also defined in advance to allow the original vision to become reality (see Fig. 15–9).

In summary, the success of the U.K. electricity and Argentinean natural gas privatizations have moved forward the debate on deregulation for energy utilities around the world. This debate and the first deregulation steps had been opened up in the United States, with its tradition of private operators, open access for pipelines, price deregulation movements at various stages of the supply chain, and the emergence of many independent

	Independent	Objective	Transparent	Efficient	Proactive
Personnel	• Respected directors • Limited basis for removal • Competitive salaries			• Small staff	
Decisional		• Decisions based on facts	• Decisions based on facts • Reasons for decisions clearly articulated		
Procedural	• Limited basis for appeal • Not limited to facts presented	• Use guidelines for major recurring decisions • Published guidelines	• Clear guidelines • Public hearings • Open access to Ente and records • Limited basis for appeals	• Limits on time for decisions • Use guidelines for recurring decisions • Published guidelines • Limited oral hearings, rely on paper hearings • Staff handling of less important decisions • Limited basis for appeal	
Financial	• Sufficient budget • Independent source of funds				
Other	• Access to outside expertise			• Uniform data reporting	• Annual report • Published performance data • Audit • Establish clear standards

Fig. 15–9 Key Regulatory Policies.

U.K. Electricity and Argentina Natural Gas		United States
• Totally deintegrated industries—competitive • Emphasis on unbundling of services and cost-based competition at all stages of supply chain • Large increases in staff efficiency • Price—caps mechanisms in distribution • Lean, transparent and fair regulators • Electric spot market created (UK) • Significant foreign capital and participation • Worldwide IPP opportunities very large and growing • Stable industry structure and regulatory framework	VS.	• Vertically integrated electric utilities • True only in parts of supply chain • Staffing inefficiencies throughout electric sector and gas LCDs • Cost-plus in distribution • Transparent, but very large and inefficient regulatory bureaucracies • True only in natural gas (spot markets) • Electric utilities and gas LCDs mostly confined within state boundaries • IPPs are a major factor in U.S., but market is maturing • Many uncertainties remain

Fig. 15–10 Contrasts Between National Industry and Regulatory Structures.

power producers. But the rest of the world, led by innovative privatizations in the United Kingdom and Argentina, is now catching up (see Fig. 15–10).

This shows that privatizations, beyond transferring expensively managed state owned assets to a more efficient private sector, can also contribute to increased competition in previously sheltered sectors. This provides incremental value to large and small customers alike through lower prices, better service, and enhanced infrastructure development.

SECTION V
CONCLUSION

CHAPTER 16

CONCLUSION

This book has been written with an integrated oil company in mind—spanning the value chain from exploration through retail marketing. For some companies, this remains a viable assumption. For others, success may be more readily achieved by focussing on a particular segment of the value chain in order to leverage their capabilities.

It is therefore opportune to end with a few comments about the value of vertical integration in the petroleum business and then to close with description of the key characteristics of the high performance petroleum company of the next century.

VERTICAL INTEGRATION REVISITED

s the oil industry moves toward the twenty-first century, vertical integration again moves into the spotlight. The ascendancy of OPEC in the decade following the 1973 embargo, and the nationalizations that occurred in its wake paved the way for national oil companies (NOCs) to emerge as powerful forces in the business. Now several of these NOCs are integrating by purchasing downstream assets.

But many of the arguments being used for reintegration rest on faulty assumptions. Although vertical integration receives most of the attention, it has been horizontal integration that has historically been used to control the oil business.

Following in the footsteps of John D. Rockefeller, the Texas Railroad Commission, and various production-sharing agreements among the majors, OPEC has been the most recent group designed to carve up production in the the industry through accords. Since 1986, however, OPEC has been fighting a losing battle. And vertical integration by the NOCs of member OPEC countries has served to make the industry less concentrated on a global basis, not more.

Increasingly the world's top producers are national oil companies, whereas in the past it was the majors. But concentration in the industry has, by most measures, been declining, not increasing. The only significant measure that indicates increasing concentration is in reserves, as the Middle East producers continue to build share. The major Western oil companies, meanwhile, have tended to deintegrate, in response to market and technological developments, with the various stages of the businesses being operated independently of each other.

The case for vertical integration has been viewed somewhat differently from the one for horizontal concentration. Rather than control, arguments for vertical integration tended to center on profitability and security of operations. Perceived benefits included:

- Reduction in profit volatility through offsetting movements at different ends of the business

- Guaranteed throughput at the production and refining stages, which lowers the risk, and therefore the cost, of capital investment
- Improved timing of investments at all stages in the chain
- Better logistics
- Allowance for creation of market outlets which might not otherwise be viewed as economically viable

However the interaction of the upstream and downstream ends of the business no longer follows conventional wisdom. The prevailing view held that changes in production profits and refining margins offset each other, and that integration provides stabilized earnings. The historical record is not so clear.

The crude business is commodity oriented. Prices respond quickly to changes in market conditions, while costs respond slowly to investment patterns and technology. Since 1973 crude prices have become more volatile by any measured, but earnings are more influenced by a few episodic shifts in the world's supply curve. In short, upstream profitability is driven by long investment cycles and are influenced by OPEC.

Refining, on the other hand, is a margin business. In the short run, crude oil price changes are reflected quickly. Over the past decade, 80% of a change in the price of crude oil has been reflected in product prices within two months. During this brief adjustment period, the upstream and downstream margins are offsetting. But to a large extent, companies at both ends could use the financial markets to offset at least a part of their exposure to large price moves on a month-by-month basis.

In the longer term, refining margins follow the investment cycle more closely than the crude price and respond to utilization levels. Estimates show that a 5% change in utilization will increase margins by a third when capacity is tight. Perhaps more important from the producing-country perspective, since 1955, upstream margins have been significantly greater than refining margins on a per-barrel basis.

The implication is that in the short run, asset ownership is not required to achieve earnings stability and that in the long run, owning refinery assets could dampen E&P earning swings—but only at the cost of owning a much lower-return assets.

The growth of broad and liquid spot, futures, options and derivatives markets has further weakened the case for vertical integration since in many major markets there is no need to control the entire value chain as a means of monetizing reserves.

FINDING A DOWNSTREAM NICHE

f integration does not make a lot of sense from a purely risk-reduction perspective, and it is not necessary to monetize reserves/production, then where does it make sense? We see integration as a focused strategy rather than a broad-based one. The reasons for integration today revert back to rationales used the industry's formative years, but the applications could differ.

Cost-saving is a major reason to integrate. Arco has used this approach on the U.S. West Coast, where it has a 19% market share in the five Western states it serves. Captive crude flows to efficiently configured refineries, serving a major market. This strategy was followed by the majors on a more global scale before 1973. It was also the cause of problems in the mid-1970's, when crude was no longer captive and refineries were not flexible enough to profitably refine alternating feedstocks.

But if a company's supply is secure (and for many NOCs it is), and if its refineries can process that feedstock at a minimum cost, then integration works.

Integration becomes an operational rather than a financial strategy. Also, shifting crude and product differentials often create major commercial opportunities which may undercut the value of tightly integrated supply chains.

Integration can also make sense when new markets must to be created and the associated infrastructure developed. Good examples of this are the moves by low gravity crude producers like Venezuela and Mexico to invest in processing facilities which can utilize these crude streams. The Shell-Pemex joint venture at Deer Park and PDVSA's investment in Citgo serve as such examples.

In today's world, the markets needing development are more likely to be gas than liquid and will often lie outside Europe and North America. Oil companies will have to think about integrating downstream on the gas value chain. New businesses could include pipelines, distribution systems and natural gas-intensive industries, such as electricity generation and petrochemicals.

In summary, integration is still a vital strategy, but its applications should be focused. The general balancing of systems on a non-integrated basis makes little sense financially, and the attendant complexity can drain too much management time. Non-strategic integration may prove to be a high-cost solution in a competitive world, leading to below-average returns.

CHARACTERISTICS OF TOMORROW'S HIGH PERFORMANCE PETROLEUM COMPANY

The high-performance oil company of the twenty-first century will have the following characteristics:

● **Builds and develops market-driving capabilities.** Whether focused on the upstream, downstream or integrated, companies need to understand the sources of their competitive advantage and build on those strengths. They also need to drive markets, not just follow them.

● **Focused like a laser on profitability and shareholder value.** Whether state or privately owned, companies will universally be more concerned with return on assets, capital employed, and operating margins. Physical measures will be replaced by financial measures. Sophisticated performance focused on management systems will replace today's measures.

● **Production decoupled from value creation.** Many companies are recognizing that exploration and production success is not sufficient to earn adequate returns; that is, true value creation comes from monetizing the resources along the value chain.

● **Flat.** Greater spans of control and fewer managerial levels (in some cases as few as four from the chairman to the technician).

● **Horizontal.** Focusing on key business processes not vertical hierarchy

● **Teamwork.** Flexibility and the flatter organizations will require unprecedented teamwork. Hierarchy will become a distant memory.

● **Changed role of the CEO and corporate headquarters.** The leading companies in the next decade may adopt very different top management and corporate structures. Shared responsibilities for the CEO, floating headquarters or multiple corporate centers, and lean corporate staffs (with many traditional functions outsourced) could all become commonplace.

● **Empowerment and accountability**. Decisions will be made lower in the organization, but this trend will be accompanied by far clearer accountability than in the past.

● **Human resource strategy as a competitive differentiator**. Those companies that figure out how to attract, develop, and retain their future leaders and how to encourage and manage diversity will have a real competitive advantage.

● **Flexibility and responsiveness**. We have all learned that forecasting the future business scenarios is fraught with difficulty. Therefore, the keys to future success are flexibility, responsiveness, speed, and agility in anticipating changes in the environment and leading the wave of response. This capability requires a new type of leadership and bureaucratic model.

● **Risk management**. Volatile markets will require proactive measurement and management of price and other financial risks.

● **Quality and environmental compliance**. Compliance measures will be important and possibly linked to financial performance.

● **Extended enterprises**. The company of the future will depend on a wide variety of service providers who will become that company's strategic partners in an "extended enterprise," linking the company's suppliers and contractors

CONCLUSION

It is impossible to boil the contents of this book into a single, neat conclusion. The task of transforming a large petroleum company into a lean, efficient, and effective enterprise is a formidable one. However, there are a few clear messages.

First, start with a strategic vision. Second, decide what capabilities are needed to successfully prosecute that strategy. Third, build those capabilities through the redesign of business processes, organizational structure, and people development. And finally, measure performance and continuously adjust to changing circumstances—inertia and conventional wisdom are the most insidious of enemies.

We have attempted to share the experiences and lessons of our many years of advising petroleum companies. We hope this book can help chart a course for your company that will ensure it is among the high performing enterprises of the next century.

APPENDIX

APPENDIX

QUALITY MANAGEMENT STANDARDS

In terms of actual quality management practices, the oil industry does not differ dramatically from other industries. The industry, like others, possesses its share of "guru" advocates, TQM advocates, ISO 9000 advocates, and industry standard (API Q1) advocates. However, the relatively greater influence of regulation on operations in the oil industry, particularly the downstream activities, causes a relatively greater tendency towards the existence of compliance programs. These compliance programs, usually in the areas of environment, health, and safety, tend to necessitate the procedure-driven, independent oversight process control systems accommodated by quality management standards such as ISO 9000 or the Responsible Care programs.

At the same time, however, the Department of Energy (DOE) seems to be moving towards a TQM model of quality management, as evidenced by DOE Order 5700.6C Quality Assurance Program. The DOE Order 5700.6C notes that almost universally, modern industry has begun to recognize that it must reach beyond traditional quality assurance methods. The basic requirements of this order are oriented towards performance, not process, placing emphasis on work results and improving the integrity of products and services provided. Hazel O'Leary, the Secretary of Energy, is a self-proclaimed advocate of TQM, as is President Bill Clinton. In summary, the quality program standards, including the award criteria for certain quality awards, which are most prevalent in the oil industry include:

- API Q1
- ISO 9000
- Malcolm Baldrige National Quality Award criteria

API Q1

The American Petroleum Institute (API), a trade group representing the nation's oil companies, began publishing its program for quality management, API Q1, in 1985. The API Q1 specification, a "QA/QC" quality management standard under the jurisdiction of the API Production Department Standardization Committee on Quality, is currently in the process of undergoing a major revision, the fifth revision to date. The fourth revision, incorporated January 1, 1992, was basically a rewrite to encompass the elements of ISO 9001. The switch to incorporate the elements of ISO 9001 was primarily driven by requests by API's customers. Revising API Q1 to more align with the ISO 9000 series standards should make it easier for API licensees to attain ISO 9000 series registration and should reduce the costs of implementation.

The committee charged with drafting the new API Q1 requirements for the fourth revision had originally voted to incorporate the ISO 9001 external quality assurance standard in its entirety, but later reconsidered when API's legal department expressed reservations. The lawyers felt that there was the potential for ISO 9000 to prohibit or infringe upon API's licensing system. API's licensing program offers product certification for companies that meet the requirements of Q1 and other criteria, which vary from product to product. The lawyers' concern was that API could not control the interpretations of ISO 9000. Given that there are legal ramifications when the API logo is on a product, control was an issue. API is in complete control of the Q1 document. Only API can make interpretations of that document.

The API Production Department has established and maintains a standardization program to facilitate the broad availability of safe, interchangeable equipment and materials for general use in the petroleum drilling and production industry, and to ensure the broad availability of proven, sound, engineering and operating practices. This program includes a provision for licensing manufacturers to use the API monogram. API will issue a license for product specification. This certificate shows that you have the capability to produce a product as specified. It does not mean you produce a great product. But if you sell a piece of pipe and it has the API monogram on it, this means that you are complying with certain product specifications. API does not certify individual users or manufacturers of product, it certifies how supplies for the oil industry were produced.

The API Q1 and ISO 9001 standards are quite similar, although not a perfect mirror image of one another. However, given that API is now certified as an ISO 9000 registrar, it is likely that differences between the two standards will remain modest.

ISO 9000

The International Organization for Standardization (ISO) is a specialized international agency for standardization, currently comprising the national standards bodies of 91 countries. The purpose of ISO is to promote the development of standardization to facilitate the international exchange of good and services and to develop cooperation in intellectual, scientific, technological, and economic activity. The results of ISO technical work are published as international standards. The American National Standards Institute (ANSI) is the member body representing the United States.

ISO is made up of some 180 technical committees. ISO Technical Committee 176 (ISO/TC176), an ISO subcommittee headquartered in Canada, was formed in 1979 to harmonize the increasing activity in quality management and quality assurance standards. Many complain that the ISO is a creature of European technical expertise, but in reality, the United States staffs many of the key technical committees, including ISO/TC176. In 1987 the ISO 9000 Series standards on quality management and quality assurance emerged as a result of ISO/TC176 efforts.

The ISO 9000 Series is a set of five individual, but related, international standards on quality management and quality assurance, none of which are specific to any particular product. ISO 9000 provides the user with guidelines for the selection and use of ISO 9001, 9002, 9003, and 9004. ISO 9001, 9002, and 9003 are quality system models for external quality assurance. These three models are actually subsets of each other. ISO 9001 is the most comprehensive, covering design, manufacturing, installation, and servicing systems. ISO 9002 covers production and installation, and ISO 9003 covers only final product inspection and test. These three models were developed for use in contractual situations such as those between a customer and supplier. ISO 9004 provides guidelines for internal use by a producer in developing its own quality system.

The choice of which model to implement depends on the company's scope of operations and the existence of externally imposed requirements to register. For example, if a company designs its own product or service, it must consider ISO 9001. Or if a company only manufactures (working off someone else's design), it may wish to consider ISO 9002. Finally, if a company neither designs nor manufactures, it may wish to consider ISO 9003. ISO 9001, 9002, and 9003 are appropriate for third-party certification. The requirement for third-party certification can either be regulatory driven or market driven.

The European Community (EC) leads in the area of regulatory driven requirements for certification. The EC has divided all products into two categories: regulated and nonregulated. Regulated products, which affect health, safety, or the environment, represent 10%–15% of all the products manufactured and sold in the EC and half of the dollar volume of all products imported from the United States. Documents called "EC Directives" describe the requirements for regulated products, including any possible requirements for ISO 9000 certification. The United States Department of Commerce maintains a listing of products regulated by the EC.

Third party registration is only one of many possible quality system verification procedures that may be specified in a given directive, ranging from a company's self-declaration of compliance through third-party registration. Registration generally involves having an accredited independent third party conduct an on-site audit of a company's operations against the requirements of the appropriate standard. Upon successful completion of this audit, the company will receive a registration certificate that identifies your quality management system as being in compliance with ISO 9001, 9002, or 9003. The company will be listed in a register maintained by the accredited third-party registration organization. The company may publicize its registration and use the third-party registrar's certification mark and the accreditation body's mark on its advertising, letterheads, and other publicity materials, but no on its products.

The certification process is improving, but remains complex due to the lack of global acceptance of member countries registrar accreditors. In the two-tiered accrediting system, each member country has one body, a registrar accreditor, that accredits other companies the ability to provide certification services. Given that circumstances are changing daily regarding certified registrars, an interested company should contact the appropriate member country's registrar accreditor. The U.S.-designated registrar accreditor is the Registrar Accreditation Board (RAB). An affiliate of ASQC, the RAB provides the latest information on certification issues. It is located in Milwaukee, Wisconsin.

The chemical and electrical/electronic industries have moved quickly to get ISO certification. DuPont, for example, has registered over 60 separate businesses worldwide. Most of the other major oil companies with petrochemical businesses have followed suit (see Fig. A–1). The chemical industry now requires most of its vendors to have certification. The certification movement is expanding from the chemical industry to the pharmaceutical industry, and is also now reaching smaller manufacturers. The ASQC is charged with registering the certifications, and the Center for Energy and Environmental Management (CEEM) in Fairfax, Virginia maintains a listing of energy companies that are certified to the ISO 9000 Series of standards.

While there are an ever-growing number of registrars for ISO 9000 certification, some have found niches for themselves in the petroleum industry. A sample of registrars serving the petroleum industry include:

- GBJD Registrars, Ontario
- OTS Quality Registrars, Houston, Texas
- Quality Systems Registrars, Herndon, Virginia
- SRI Quality System Registrars, Wexford, Pennsylvania
- Warnock Hersey Professional Services, LaSalle, Quebec

DoD, NASA, and the Big-three automakers are phasing out some of their own quality standards and adopting ISO 9000 instead as a primary

		Location	2 digit SIC	ISO Standard
AMOCO–Amoco Chemical Company	Manuf. of purified Terephthalic Acid	Wando, TX	2800	9002
	Production of Paraxylene, styrene, etc.	Texas City, TX	2800	9002
	Production of polybutene	Whiting, IN	2800	9002
	Manufacture of Purified Terephtalic Acid	Decatur, IL	2800	9002
	Manufacture of olefins and polymers	Alvin, TX	2800	9002
	Manufacture of olefins and polymers	Houston, TX	2800	9002
	Production of isophthalic acid, etc.	Joliet, IL	2800	9002
	Manufacture of paraxylene	Decatur, IL	2800	9002
Amoco Oil Company	Analytical laboratory testing services	Whiting, IN	8700	9002
Amoco Performance Products, Inc	Manufacture and supply of Xydar resins, etc.	Augusta, GA	2800	9002
	Manufacture of engineering thermoplastic resins, etc.	Marietta, GA	2800	9002
	Manufacture of Thornel polyacrylonitrile carbon fibers	Piedmont, SC	2800	9002
Amoco Petro. Additives Co.	Manufacture of lubricating and fuel additives, etc.	WoodRiver, IL	2800	9002
Amoco Torlon Products, Inc.	Design, manuf. and supply of engineering polymers	Atlanta, GA	2800	9001
CHEVRON–Chevron Chemical Company	Manufacture of ethylene, propylene, acetylene, etc.	Bautown, TX	2800	9002
	Sales, Supply, distrib. of ethylene, propylene, etc.	Houston, TX	2800	9002
	Blending of lubricating oil additives, etc.	BelleChasse, LA	2900	9002
	Manuf. of high purity styrene monomer etc.	St James, LA	2800	9002
EXXON–Exxon Chemical Company	Manufacture of polyalkenes	Baytown, TX	2800	9002
	Manuf/ supply of hydrocarbon-based polymeric resins	BatonRouge, TX	2800	9002
	Manuf/supply of butyl polymers and Vistanex etc.	Baytown, TX	2800	9002
	Manuf/supply of Vistalon rubber	BatonRouge, TX	2800	9002
	Design/develop. of process controls for Exxon plants	Baytown, TX	7300	9001

Fig. A–1 ISO 9000 Certifications. Source: CEEM Information Services, Fairfax, Virginia.

standard, then supplementing this with standards tailored to their particular supplier base. Many industries are embracing ISO 9000 and supplementing it with their own requirements. The aerospace and defense industries, for example, have put together NADCAP (National Aerospace &

EXXON–Exxon Chemical (con't) Company	Manuf/supply of halobutyl rubber etc.	BatonRouge, TX	2800	9002
	Inspection/measure/test provisions, Baton Rouge Plnt.	BatonRouge, TX	7300	9001
	Generation of technical data of elastometic samples	Linden, NJ	7300	9001
	Provision of analytical services for Exxon chemicals	Bayonne, NJ	7300	9002
	Provision of analytical services for Exxon chemicals	Linden, NJ	7300	9002
	Manuf/supply of oil additives	Linden, NJ	2800	9002
	Design/development of fuel/lubricant addtives	Linden, NJ	2800	9001
	Analyzer data and supporting technical services	BatonRouge, LA	2800	9002
	Marketing of polymers for the tire industry	Aakron, OH	2800	9002
MOBIL–Mobil Chemical Company	Manufacture/supply of oriented polypropylene film	LaGrange, GA	3000	9002
	Manuf. of commercially oriented polypropylene films	Macedon, NY	3000	9002
	Manuf/supply of biaxially oriented polypropylene film	Shawnee, OK	3000	9002
	Manuf. of commercially oriented polypropylene films	Stratford, CT	3000	9002
	Planning of production capacity for the N.A. films div.	Pittsford, NY	8900	9002
TEXACO–Texaco Chemical Company	Manuf. of TEXOX polyols and functional fluids, etc.	Conroe, TX	2800	9002
	Contract review for various chemical products	Houston, TX	2800	9002
	Manuf of Ethylene, Propylene, Benzene, etc	PortArthur, TX	2800	9002
	Manuf of Ethylene, Propylene, etc	PortNeches, TX	2800	9002
	Manufacture of specialty chemicals	Austin, TX	2800	9002
	Manufacture of 1,3 Butadiene, MTBE, etc.	PortNeches, TX	2800	9002
SHELL–Shell Chemical Company	Manuf/Supply of polybtylene	Hahnville, LA	2800	9002
	Manuf of ethylene oxide catalyst	Martinez, CA	2800	9002
	Manuf/Distrib. of ethylene, propylene and butadiene	Norco, LA	2800	9002

Fig. A–1 ISO 9000 Certifications. Source: CEEM Information Services, Fairfax, Virginia.

Defense Company Accreditation Program). The member companies put forth examples of their "best practices" in the area of quality management, and then a committee selects the best elements for codification into the NADCAP program.

SHELL–Shell Chemical -(con't) Company	Manuf/supply of commercial Kraton thermoplastics	Belpre, OH	3000	9002
	Provision of sourcing and procurement of raw materials	Houston, TX	7300	9002
Shell Oil Company	Provision of sourcing and monitoring warehouse	Houston, TX	7300	9002
	Manuf/supply of lubricating oils and greases	Carson, CA	2900	9002
	Manuf/supply of lubricating oils and greases	DeerPark, TX	2900	9002
	Manuf/supply of lubricating oils and greases	Metairie, LA	2900	9002
	Manuf/distrib of resin products	DeerPark, TX	2800	9002
	Manuf/distrib of resin products	Argo, IL	2800	9002
	Services for inspecting, chartering vessels	Houston, TX	2800	9002
ARCO–Arco Products Co.	Production of calcined coke	Blaine, WA	2900	9002
Arco Chemical Company	Customer support, materials purchase activities	Newtown Sq.PA	2800	9002
	Manuf of propylene oxide and glycols, arconates, etc.	Pasadena, TX	2800	9002
BP–BP Chemicals	Sale of acetyls, solvents, petrochemicals, etc.	Hackettstown, NJ	2800	9002
	Sales/Marketing of acrylonitrile, acetonitrile, etc	Cleveland, OH	2800	9002
	Sales/Marketing of acrylonitrile, acetonitrile, etc	Lima, OH	2800	9002
	Sales/Marketing of acrylonitrile, acetonitrile, etc	PortLavaca, TX	2800	9002
	No certificates on file			
CONOCO	No certificates on file			
USX-MARATHON	1 certificate on file			
PHILLIPS PETROL.	2 certificates for chemicals on file			
ASHLAND OIL	No certificates on file			
COASTAL	No certificates on file			
UNOCAL	3 certificates for chemicals on file			
OCCIDENTAL				

Fig. A–1 ISO 9000 Certifications. Source: CEEM Information Services, Fairfax, Virginia.

The whole point of ISO 9000 is to prove that a company does what it says it does. Companies are able to demonstrate that quality programs are functioning, that their manufacturing process is reliable and consistent, and that they keep good records. ISO 9000 does not guarantee a quality product, however. A production facility that is ISO certified and that keeps good records, could still be consistently turning out a lousy television. ISO 9000 does not specify what particular industry standard must be used. It could be NQA-1 if one's company is in the nuclear power business, or it can be that company's own proprietary standard. So long as the company seeking certification is consistent in application (and of course, record-keeping) it is meeting the requirements of ISO 9000. ISO 9000 is a business management tool that lets allows one to trace why a product was poorly produced and fix the problem.

ASQC drafts U.S.-equivalent standards to the ISO standards. In addition to translating into American English from British English, they employ a slightly different terminology:

- ISO 9000 = Q90
- ISO 9001 = Q91
- ISO 9002 = Q92
- ISO 9003 = Q93
- ISO 9004 = Q94

The ISO 9000 Series is reviewed about every five years. The first revision to the base standards 9000–9004 began in 1992. Approvals from the member countries are anticipated to be received and the updates complete by no later than mid-year 1995. The draft versions of the revisions being examined have been released as Draft International Standards (DIS). The revisions reflect a clear move towards encompassing continuous improvement themes as well as requirements pertaining to safety and the environment.

MBNQA

The Malcolm Baldrige National Quality Improvement Act, signed by President Ronald Reagan on August 20, 1987, originated the Malcolm Baldrige National Quality Award (MBNQA). The Act, named after a former U.S. Secretary of Commerce, called for the creation of a national quality award and the development of guidelines and criteria that organizations could use to evaluate their quality improvement efforts. Awards were to be given in three categories — manufacturing, service, and small business — with no more than two awards per year. See Figure A–2 for winners through 1993. The legislation gave favorable mention to a number of management principles and tools: worker involvement, strategic quality planning, statistical process control, management-led and customer -oriented programs. But the Act said little about the award's scoring

**MALCOLM BALDRIGE NATIONAL QUALITY AWARD
1993 WINNERS**

COMPANY	QUALITY CONTACT	PUBLIC AFFAIRS CONTACT
EASTMAN CHEMICAL COMPANY *Manufacturing*	Robert C. Joines Vice President, Quality P.O. Box 511 Kingsport, TN 37662 Tele: 615-229-5658 Fax: 615-229-1351	Rodney D. Irvin, Manager Media Relations P.O. Box 511, Stone East Building 2550 East Stone Drive Kingsport, TN 37662 Tele: 615-229-4008 Fax: 615-229-1008
AMES RUBBER CORPORATION *Small Business*	Charles A. Roberts Vice President, Total Quality 23-47 Ames Boulevard Hamburg, NJ 07419 Tele: 201-209-3200 Fax: 201-827-8893	Same As Quality Contact

**MALCOLM BALDRIGE NATIONAL QUALITY AWARD
PREVIOUS WINNERS**

COMPANY	QUALITY CONTACT	PUBLIC AFFAIRS CONTACT
AT&T NETWORK SYSTEMS GROUP **Transmission Systems Business Unit - 1992** *Manufacturing*	Louis E. Monteforte, Director Transmission Quality Planning 475 South Street, Room 2W-44 Morristown, NJ 07962-1976 Tele: 201-606-2488 Fax: 201-606-3363 or Quality Hotline: 1-800-682-7759	Jill Christen-Betline Public Relations Specialist Transmissions Systems 475 South Street, Room 2W-1 Morristown, NJ 07962-1976 Tele: 201-606-2534 Fax: 201-606-3307
CADILLAC MOTOR CAR COMPANY - 1990 *Manufacturing*	Joseph R. Bransky, Director Quality and Reliability General Motors Corporation General Motors Building, Rm. 6-162 Detroit, MI 48202 Tele: 313-556-9050 Fax: 313-974-6899	John F. Maciarz, Director Corporate Communications & NAO Public Affairs General Motors Corporation General Motors Building, Rm. 11-261 Detroit, MI 48202 Tele: 313-556-2034 Fax: 313-974-4485

Fig. A-2 Malcolm Baldrige National Quality Award 1993 Winners. Sources: Malcolm Baldrige National Quality Department, National Institute of Standards and Technology (NIST).

system, judging process, or criteria for evaluation. It was left to the National Bureau of Standards (known today as the National Institute of Standards and Technology, or NIST) to work out the details.

Collaborating closely with industry experts, NIST produced the seven category, 1,000 point scoring system and the three level judging process that are still used today. The criteria and subcategories have evolved over time. Companies submit applications of up to 75 pages (up to 50 pages for small businesses) describing their quality practices and performance in each of the seven

FEDERAL EXPRESS CORPORATION - 1990 *Service*	**Quality Questions & Information** Jean Ward-Jones, Manager Quality Education & Administration P.O. Box 727 Memphis, TN 38194-2142 Tele: 901-395-4539 Fax: 901-395-4641	**Quality Speaker Service** Sally Davenport, Senior Specialist Public Relations 2005 Corporate Avenue Memphis, TN 38132 Tele: 901-395-3466 Fax: 901-395-4928 **Reporters & Editors Only** Shirley Finley, Senior Specialist Media Relations 2005 Corporate Avenue Memphis, TN 38132 Tele: 901-395-3463 Fax: 901-346-1013 or 395-4928
GLOBE METALLURGICAL, INC. - 1988 *Small Business*	Norman Jennings, Quality Director P.O. Box 157 Beverly, OH 45715 Tele: 614-984-2361 Fax: 614-984-8695	Sherryl Hennessey Public Relations P.O. Box 157 Beverly, OH 45715 Tele: 614-984-2361 Fax: 614-984-8695
GRANITE ROCK COMPANY - 1992 *Small Business*	Bruce W. Woolpert President and CEO P.O. Box 50001 Watsonville, CA 95077-5001 Tele: 408-761-2300 Fax: 408-724-3484	Greg Diehl, Manager Marketing Services P.O. Box 50001 Watsonville, CA 95077-5001 Tele: 408-724-5611 Fax: 408-724-3484
IBM ROCHESTER - 1990 *Manufacturing*	IBM Rochester Center for Excellence 3605 Highway 52 North Rochester, MN 55901-7829 Tele: 507-286-5000 (Hotline) Fax: 507-286-5010	Richard Ulland Staff Communications Specialist 3605 Highway 52 North, 70C/107-1 Rochester, MN 55901-7829 Tele: 507-253-4340 Fax: 507-253-7175
MARLOW INDUSTRIES - 1991 *Small Business*	Joy Janco Baldrige Activities Coordinator 10451 Vista Park Road Dallas, TX 75238-1645 Tele: 214-342-4293 Fax: 214-341-5212	Same As Quality Contact

Fig. A–2 Malcolm Baldrige National Quality Award 1993 Winners. Sources: Malcolm Baldrige National Quality Department, National Institute of Standards and Technology (NIST).

required areas. The applications are graded by teams of trained examiners. The Baldrige judges, recognized quality experts who come from industry, academia, and consulting firms, choose a small set of high-scoring applications for visits. A team of senior examiners and examiners visits each company for at least several days, conducting interviews and checking documents. The judges then meet a final time to review the top applicants and to select winners.

MILLIKEN & COMPANY - 1989 *Manufacturing*	Dr. Patrick C. Bowie Vice President, Quality P.O. Box 1926, M-186 Spartanburg, SC 29304 Tele: 803-573-2003 Fax: 803-573-2505 Tours: Sandra Howell Tele: 803-573-1988	Richard Dillard Director of Public Affairs P.O. Box 1926, MS-285 Spartanburg, SC 29304 Tele: 803-573-2546 Fax: 803-573-2100
MOTOROLA, INC. - 1988 *Manufacturing*	Richard Buetow Senior Vice President & Director of Quality 1303 East Algonquin Road Schaumburg, IL 60196 Tele: 708-576-5516 Fax: 708-538-2663	Margot Brown Director, Media Relations 1303 East Algonquin Road Schaumburg, IL 60196 Tele: 708-576-5304 Fax: 708-576-7653
SOLECTRON CORPORATION - 1991 *Manufacturing*	Margaret Smith Marketing Program Specialist 847 Gibraitar Drive Milpitas, CA 95035 Tele: 408-956-6768 Fax: 408-956-6056	**Media Requests Only** Jeffrey F. Cox, Manager Corporate & Marketing Communications 847 Gibraiter Drive Milpitas, CA 95035 Tele: 408-956-6688 Fax: 408-956-6056
TEXAS INSTRUMENTS INCORPORATED **Defense Systems & Electronics Group - 1992** *Manufacturing*	Mike Cooney, Vice President Quality Assurance P.O. Box 660246, M/S 3124 Dallas, TX 75266 Tele: 214-480-4800 Fax: 214-480-4880	Tony Geishauser, Manager Media Relations P.O. Box 660246, M/S 3134 Dallas, TX 75266 Tele: 214-480-1417 Fax: 214-480-3281
THE RITZ-CARLTON HOTEL COMPANY - 1992 *Service*	Patrick Mene Corporate Director of Quality 3414 Peachtree Road, N.E., Suite 300 Atlanta, GA 30326 Tele: 404-237-5500 Fax: 404-261-0119	Karon Cullen Corporate Director of Public Relations 3414 Peachtree Rd., N.E., Suite 300 Atlanta, GA 30326 Tele: 404-237-5500 Fax: 404-237-9643
WALLACE COMPANY - 1990 *Small Business*	Assets of the Wallace Company have been acquired by the industrial supply division of another company.	

Fig. A–2 Malcolm Baldrige National Quality Award 1993 Winners. Sources: Malcolm Baldrige National Quality Department, National Institute of Standards and Technology (NIST).

The Malcolm Baldrige National Quality Award (MBNQA) is essentially the sole formalized standard generally accepted as being a TQM standard. The merits of MBNQA serving as this standard are reflected in the numerous attempts and references made by other quality management standards, such

WESTINGHOUSE ELECTRIC CORP. **Commercial Nuclear Fuel Division - 1988** *Manufacturing*	Same As Public Affairs Contact	Carl Arendt, Manager Communications Productivity and Quality Center P.O. Box 160 Pittsburgh, PA 15230-0160 Tele: 412-778-5008 Fax: 412-778-5153
XEROX CORPORATION **Business Products & Systems - 1989** *Manufacturing*	John G. Lawrence, Manager Quality Communications Office 1387 Fairport Road, Building 1100 Fairport, NY 14450 Tele: 716-383-7502 Fax: 716-383-7517	Samuel M. Malone, Jr., Manager Quality Communications 1387 Fairport Road, Building 1100 Fairport, NY 14450 Tele: 716-383-7534 Fax: 716-383-7517
ZYTEC CORPORATION - 1991 *Manufacturing*	Karen Scheldroup Baldrige Office 7575 Market Place Drive Eden Prairie, MN 55344 Tele: 612-941-1100 x104 Fax: 612-829-1837	Same As Quality Contact

Fig. A–2 Malcolm Baldrige National Quality Award 1993 Winners. Sources: Malcolm Baldrige National Quality Department, National Institute of Standards and Technology (NIST).

as ISO 9000, to evolve to the quality management system described by MBNQA. As indicated later, the virtues of the MBNQA criteria itself, not the award process, are generally recognized by the oil industry. As noted earlier, TQM does not approach quality management like many of the formal standards for quality management do, such as ISO 9000. For a summary of the differences between MBNQA, a reasonable proxy for TQM, and ISO 9000, refer to Figure A–3.

It is difficult to assess the percentage of oil companies that have applied for the award, given that it is the strict policy of the MBNQA to not release any information on either the companies that have applied for the award or the industry sectors represented in the selection process. As discussed in the next section, the oil industry continues to show interest in the continuous improvement, or total quality management, concepts underlying the Award criteria. Although Booz·Allen research indicates that few oil companies have applied for the award, there are several individuals from the oil industry who have served in years past or are currently serving on the MBNQA Board of Directors, including Amoco Chemical, Mobil Chemical, and Texaco. Broader information concerning the number of applications received within the manufacturing category, a relevant category for the oil industry, is provided (see Fig. A–4).

	MBNQA	ISO 9000
Focus	Competitiveness—criteria reflect two key competitiveness thrusts: 1) delivery of ever-improving value to customers, and 2) improvement of overall company operational performance.	Conformity to practices specified in the registrant's own quality system.
Purpose	Educational—to encourage sharing of competitiveness learning and to "drive" this learning nationally. It fulfills this purpose by: 1) promoting awareness of quality as an important element in competitiveness, 2) recognizing companies for successful quality strategies, and 3) fostering information sharing of lessons learned.	To provide a common basis for assuring buyers that specific practices, including documentation, are in conformance with the providers' stated quality systems. Some organizations use the ISO 9000 standards to bring basic process discipline to their operations.
Meaning of "Quality	Customer-driven quality—addressing total purchase, ownership, relationship experience—concerned with all factors that matter to customers. Conformity issues are included in criteria, under Process Management, which addresses other key operational requirements.	Conformity of specified operations to documented requirements.
Improvement Results	Depends heavily on results—"Results" are a composite of competitiveness factors: customer-related, employee-related, product and service quality, and overall productivity. "Management by fact," tied to results, is a core value. Trends (improvement) and levels (comparisons to competitors and best performance) are taken into account. Results play dual role: 1) representing business improvement indicators needed to demonstrate a successful quality strategy, and 2) representing indicators that drive improvement.	Does not assess outcome-oriented results or improvement trends. Does not require demonstration of high quality, improving quality, efficient operations or similar levels of quality among registered companies.
Role in the Marketplace	A form of recognition. Despite its heavy reliance on results, it is not intended to be a product endorsement, registration or certification. Award winners may publicize and advertise their recognition and must share quality strategies with other U.S. organizations. The winners' role is public education and inducement for others to improve. Award winners adhere to a voluntary advertising guideline that prohibits attributing their awards to their products	Provides customers with assurances that a registered supplier has a documented quality system and follows it. In some cases, registration will reduce the number of independent audits otherwise conducted by customers. While some registrars encourage advertising registration as a market advantage, some also prohibit advertising that registration signifies a product evaluation or high quality. ISO 9000 registration does not translate meaningfully into a Baldrige Award assessment score.
Nature of the Assessment	Involve s a four-stage review conducted by volunteer private-sector Board of Examiners. Applications reviewed by five to 15 board members, depending on application progress. Final contenders receive site visits (two to five days) by a team of six to eight examiners. Focus is the customer and the marketplace. Evidence of pervasive improvement, backed by results, must be in place, improvement includes customer-related and operations-related factors. It may include relevant financial indicators provided they are tied to the other indicators. Conformity and documentation are addressed as part of process management. Assessment is not an audit or conformity assessment.	Evaluates organization's quality manual and working documents, a site audit to ensure conformance to stated practices and periodic re-audits after registration. Focus is on documentation of a quality system and on conformity to that documentation.
Feedback	Applicants receive feedback covering 28 items in seven categories. Feedback is diagnostic, highlighting strengths and areas for improvement in overall competitiveness management system. Three scoring dimensions: approach, deployment and results	Audit feedback covers discrepancies and findings related to practices and documentation. Feedback takes the form of major and minor nonconformities. Assesses organization's documented quality system requirements and deployment of these requirements.

Fig. A–3 MBNQA Criteria Compared to ISO 9000 Criteria. Source: Reimann, Curt W. and Hertz, Harry S., The Malcolm Baldrige National Quality Award and ISO 9000 Registration, ASTM Standardization News, November 1993, pp. 42–53.

Criteria Improvement	Criteria booklet is revised annually to capture lessons learned from each cycle. Since 1988, the document has undergone six cycles of improvement, becoming more focused on business management. Major changes include a greater results orientation; more emphasis on speed, competitiveness, productivity, improved integration of quality and other business management requirements; greater emphasis on human resource development; and better accommodation to service organizations' requirements.	Revisions of ISO 9001, 9002 and 9003 will be issued in 1994, with a focus on clarification of the 1987 documents, themselves based on the first commercial quality system standard, BS5750, developed in 1979 (Sawin and Hutchens, 1991). The roots of BS 5750 trace to MIL-Q-9858A, established by the U.S. Department of Defense in 1959. Since 1987, additional guidance documents have been and are being developed.
Responsibility for Information Sharing	Award winners are required to share nonproprietary information on their successful quality strategies with other U.S. organizations.	Registrants have no obligation to share information with others.
Service Quality	A principal concern in guiding criteria evolution has been compatibility with service excellence. Criteria and supporting information are evaluated to improve compatibility with requirements for service excellence. Criteria are relevant to service organizations. The most important "process" item (Customer Relationship management) is a principal concern for manufacturers which seek competitive advantage via service.	ISO 9000 standards are directed towards the demonstration of a supplier's capability to control the processes that determine the acceptability of product or service, including design processes in ISO 9001, ISO 9000 standards are more oriented toward repetitive processes, without an equivalent focus on critical service quality issues such as relationship management and human resource development.
Scope of Coverage	Criteria address all operations and processes of all work-units to improve-overall company productivity, responsiveness, effectiveness and quality. Approach offers wide latitude in developing customer-focused cost reduction strategies, such as re-engineering of business processes. Due to broader nature of the Baldrige criteria and assessment, a rigorous audit of a printed quality manual and compliance with its procedures do not occur during an assessment.	ISO 9001 registration covers only design / development, production, installation and servicing. Registration covers parts of several items in the Baldrige Award criteria (primarily, parts of Management of Process Quality). ISO 9001 requirements address less than 10 percent of the scope of the Baldrige criteria, and do not fully address any of the 28 criteria items. All ISO 9001 requirements are within the scope of the Baldrige Award.
Documentation Requirement	Criteria do not spell out ongoing documentation requirement. Criteria imply that documentation should be tailored to fit requirements and circumstances, including internal, contractual or regulatory requirements. Some analysts confuse application report with an ongoing documentation requirement. Assessment relies on evidence and data, but this does not define or prescribe a documentation system.	Documentation is central audit requirement. Documentation requirements are ongoing, meaning that documents are permanent part of the quality system needed to maintain registration.
Self-Assessment	Principal use of criteria is in self-assessment of improvement practices. Inclusion of a scoring system and evaluation factors allows companies to chart their own progress. Some companies correlate their progress in Baldrige criteria self-assessment with change in financial indicators.	ISO 9000 standards are used primarily in "contractual situation" or other external audits. Additional registrar-developed audit checklists define actual criteria / requirements for registration. Registration by external assessor is needed to fulfill most contractual requirements (i.e., self-assessment is generally not accepted). Aside from benefits of self-assessment while pursuing registration, it is not clear that ISO 9000 self-assessment after registration leads to operational improvement because standards do not address continuous improvement or competitiveness factors.

Fig. A–3 MBNQA Criteria Compared to ISO 9000 Criteria. Source: Reimann, Curt W. and Hertz, Harry S., The Malcolm Baldrige National Quality Award and ISO 9000 Registration, ASTM Standardization News, November 1993, pp. 42–53.

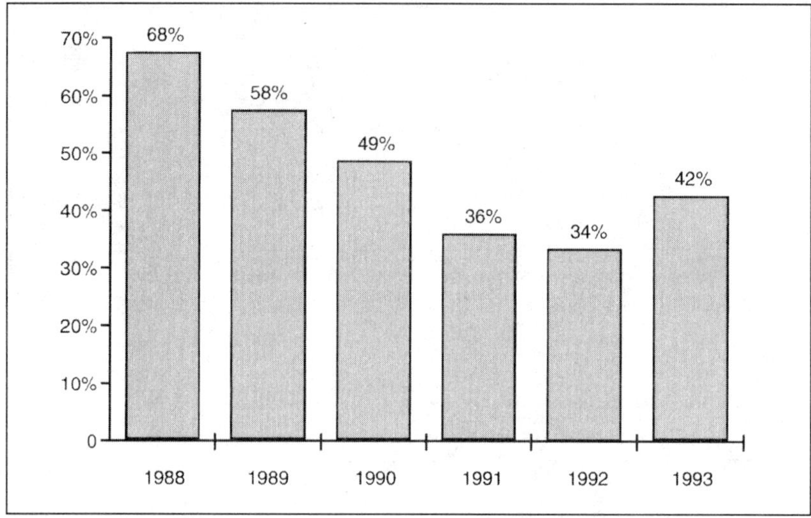

Fig. A–4 Manufacturing Interest in Applying For MBNQA.

INDEX

A

B

C

D

E

H

I

J

K

L

M

N

O

P

S

T

U

V

W

Y